Mongooses of the World

Published by
Whittles Publishing Ltd.,
Dunbeath,
Caithness, KW6 6EG,
Scotland, UK

www.whittlespublishing.com

ISBN 978-184995-435-8

Front cover: Yellow mongoose (© Emmanuel Do Linh San)
Back cover: Common dwarf mongoose (© Julie Kern)

Mongooses of the World

Andrew Jennings & Géraldine Veron

Whittles Publishing

We would like to dedicate this book to our families and friends, and to all the people working hard to study and conserve mongooses

Contents

Preface

Mongooses are a remarkable and fascinating group of small carnivores: 34 species are spread across the two continents of Africa and Asia, on which they live within a wide variety of habitats, from open savannah to dense rainforest, and display an amazing diversity in social behaviour, with both solitary and group-living species. Yet this family of mongooses is one of the least known group of carnivores. Apart from a few species that live in Africa, such as the endearing and charismatic meerkats, most are rarely seen in the wild, and several have never been studied in the field.

Species extinction is a topical subject in today's world, and much attention is focused on large carnivores, such as tigers, wolves and bears, due to their high sensitivity to human-caused threats: the destruction of native habitats, for example, and hunting for their fur, meat or other body parts. But whereas these large carnivore species have been well covered in numerous books, television documentaries and cinematic films, most of the smaller carnivores, including the mongooses, have generally been very poorly represented in the written and visual media, and have not been given the attention that they rightly deserve – mongooses are just as attractive, captivating and intriguing as their larger cousins, and face many of the same conservation threats.

Meerkats and banded mongooses are well known and extremely popular with the general public, but there is a general lack of knowledge and awareness about the rest of the mongoose species. Through reading this book, you will discover the wonderful diversity of the mongoose family, and hopefully be inspired to take a deep interest in their continued survival in the wild.

The first section of this book gives a comprehensive overview of all aspects of the natural history of mongooses, including their taxonomy and systematics, morphology, diet and foraging behaviour, breeding and social organisation, their relationships with humans and aspects of their conservation. The second section comprises detailed species accounts on all 34 mongoose species and includes distribution maps and up-to-date information on their ecology and conservation status.

About the Authors

Dr. Andy Jennings is a field researcher who has worked on numerous small carnivore projects around the world, including several studies on mongooses. Professor Géraldine Veron is a researcher in the *Institut de Systématique, Evolution, Biodiversité* and curator of Mammals in the *Museum National d'Histoire Naturelle*, Paris, and has been studying the evolution and ecology of mongooses and other small carnivores for over 20 years. Together, they completed the first radio-telemetry project on the short-tailed mongoose in Southeast Asia, and have published several scientific papers and book chapters on mongooses.

Acknowledgements

We give many thanks to all the contributors who provided photographs and very generously gave permission for us to use them in this book: Emmanuel Do Linh San, Kalyan Varma, Julie Kern, Laila Bahaa-el-din, Chris and Mathilde Stuart, Claude Fischer, Manuel Ruedi, Vijay Anand Ismavel, Jarryd Streicher, Laurent Vallotton, Nadine Cronk, Francesco Rovero, Luke Hunter, Lars Werdelin, Jayasilan Mohd-Azlan, Kevin Schafer, Lourens Swanepoel, Jason Woolgar, Baptiste Mulot and Geraldine Laval from ZooParc de Beauval, Divya Mudappa, Ondřej Machač, Cécile Callou, Linda Evans, Devcharan Jathanna, Thierry Aebischer and the Chinko Research Team, Gboja Mariano Houngbedji and Chrystelle Dakpogan Houngbedji (and the Organisation pour le Développement Durable et la Biodiversité, Bénin), Gonçalo Curveira Santos, and Andrew Hearn. We would also like to thank the many people who offered to contribute photographs for this book, or provided contacts, support, and encouragement to this project: Dez Fernandez, Vivien Louppe, Gopalasamy Reuben Clements, Andre Pittet, Peter Taylor, Mathias D'Haen, Meaghan Evans, Thomas N. E. Gray, G.A. Punjabi, Ulrike Streicher, Riddhika Kalle, Tania L.F. Bird, Luke Verburgt, Colleens Downs, Torsten Bohm, Bogdan Cristescu, Andy Radford, Amy Dunham, A. Cole Burton, Claudio Sillero-Zubiri, Milan Korinek, Thomas Engel, Samundra Ambuhang Subba (and WWF Nepal), and Jedediah Brodie.

Special thanks go to Luke Hunter, Priscilla Barrett and Jenny Campbell for the permission to reproduce artwork from *A Field Guide to Carnivores of the World*, second edition, by Luke Hunter and Priscilla Barrett, Bloomsbury Wildlife (UK), an imprint of Bloomsbury Publishing Plc.

We thank Junaidi Payne for permission to reproduce artwork from *A Field Guide to the Mammals of Borneo*, by J. Payne, C.M. Francis and K. Phillipps, Sabah Society and WWF Malaysia; Jonathan Kingdon for permission to reproduce artwork from *East African Mammals. An Atlas of Evolution in Africa* by J. Kingdon, Academic Press; and the Bombay Natural History Society for permission to reproduce artwork from *The Book of Indian Animals* by S.H. Prater.

We thank Lynne Cullen for helping to improve the quality of several photographs.

We would like to give special thanks to Keith Whittles for his enthusiastic encouragement in initiating this book project. And finally, we give thanks to all the people and institutions that made it possible for us to conduct research on mongooses.

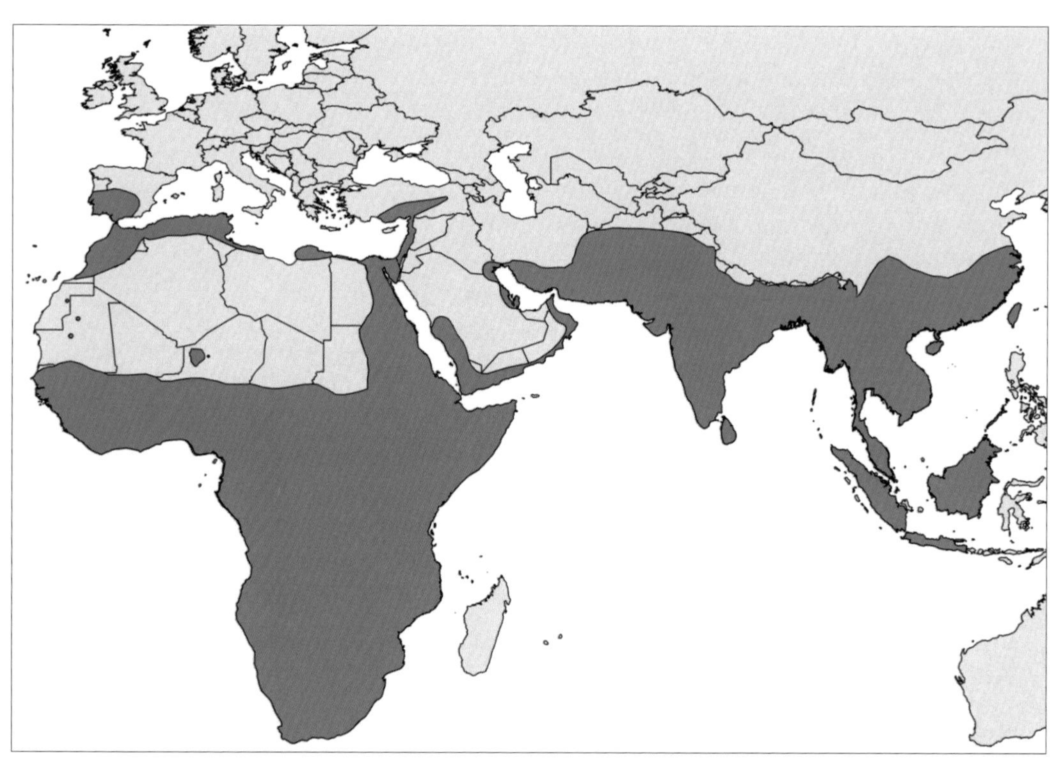

A world distribution map of mongoose species, which shows that they occur in Africa and Asia, as well as southern Europe and parts of the Middle East (© Andrew P. Jennings and Géraldine Veron)

The natural history of mongooses

1. What is a mongoose?

Mongooses are small carnivores, weighing less than 5kg and found only in the Old World. There are 25 species that live in Africa, and nine in Asia; a few mongooses also occur in southern Europe and the Middle East, and one mongoose species, in particular, has been introduced by humans to many other parts of the world. Mongooses range as far north as southern Europe and South China, and extend southwards to the southern tip of Africa and the island of Java, but they are mostly absent from the desert regions of the Sahara and Arabian Peninsula. Mongooses can be found over a wide altitudinal range, from sea level to the upper reaches of mountains, and they live within a wide variety of habitats and climates, from hot, dry, open savannahs to humid, dense rainforests. Some species are mainly solitary, whereas others live in social groups.

Not all mongooses have a common name that includes the word 'mongoose' – there are four African species of cusimanse, and of course, there is the meerkat, the best known and most popular mongoose. The name 'mongoose' is likely derived from an Indian word, such as the Marathi name *mungūs*. The Portuguese borrowed this term to commonly call these species a *mangus*, and the French subsequently named it *mangouste*. Various English-language publications spelled the name as *mungoes*, *mongoos*, *mungoose*, *mungoos* and *mangoust*, before settling on the standard spelling 'mongoose' in the 19th century. As this spelling contains the syllable '-goose', some people pluralize mongoose as 'mongeese', and the Oxford English Dictionary lists this as an irregular plural that is occasionally used. However, the standard practice in English for imported foreign words is to use either the foreign language's own pluralization (if this is well known) or the English rule of adding '–s' as a suffix. As very few English-speakers know how Marathi or other Indian-derived words are pluralized, the plural form 'mongooses' is normally used.

Taxonomy and systematics

Taxonomy is the science of describing and naming biological organisms, which are then grouped together based on shared characteristics. A species is the most basic taxonomic unit, and is often defined as a group of similar organisms that are capable of exchanging genes or interbreeding; however, since the evolution of organisms is an on-going and complex process, it is often not that simple to determine what is a species. All species are given a two-part Latin name: the first part is the genus to which the species belongs, and the second is called the specific name; for example, the scientific name of the yellow mongoose is *Cynictis penicillata*.

Systematics is the study of the diversification of organisms, both past and present, and their relationships through time. These affinities are based on shared features and are visualised as evolutionary or phylogenetic 'trees', which are branching diagrams that show the inferred relationships between species.

Since the early 19th century, naturalists have classified mongooses using external characteristics, such as fur colour, and anatomical features, especially the structure of the teeth and bones. Different scientists, however, interpreted these physical features in different ways, which has resulted in several classification schemes. Over the past few decades, geneticists and evolutionary scientists have been able to extract and sequence DNA from living or dead animals, and by comparing different genes, they can identify species and determine their relationships with each other in a much more objective way than simply by using physical characteristics.

Mongooses have been grouped together with other carnivores (felids, canids, bears, seals, etc.) within the order Carnivora, as they all share a distinctive dental feature: the carnassial teeth. These two teeth (the fourth upper premolar and first lower molar) have sharp cutting edges and are located on either side of the jaw. They work together like a pair of scissors and enable carnivores to slice meat – domestic pet owners will no doubt have noticed that their cat eats meaty food using the side of its mouth. This dental adaptation has allowed carnivores to include animal flesh in their diet, and cats, in particular, have

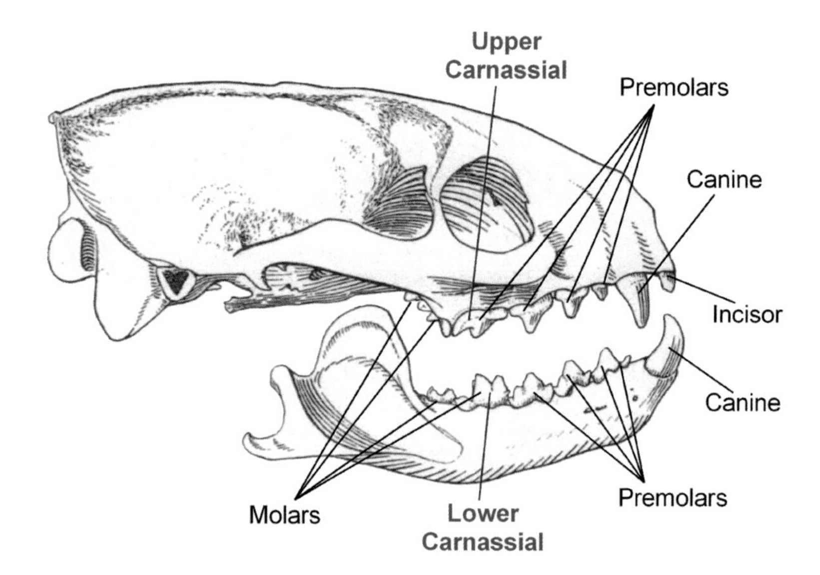

An Egyptian mongoose skull showing the various types of teeth: incisors, canines, premolars, and molars. The carnassial teeth (the fourth upper premolar and the first lower molar) are typical of carnivores – their sharp blades function like a pair of scissors and enable mongooses to slice meat (© Andrew P. Jennings and Géraldine Veron)

well developed carnassial teeth and are almost exclusively meat-eaters. Some carnivore species have additional dental adaptations that also enable them to eat other types of food; for instance, canids can munch a wide variety of plants and vegetables, and hyaenas can crush hard bones, whereas others, such as the giant panda and red panda, are virtually vegetarian, eating primarily bamboo. Mongooses are mainly animal eaters, although their diet varies greatly between different species and in different places.

Currently, the order Carnivora comprises over 280 species, ranging in size from the tiny least weasel to the large polar bear and elephant seal. These species are arranged

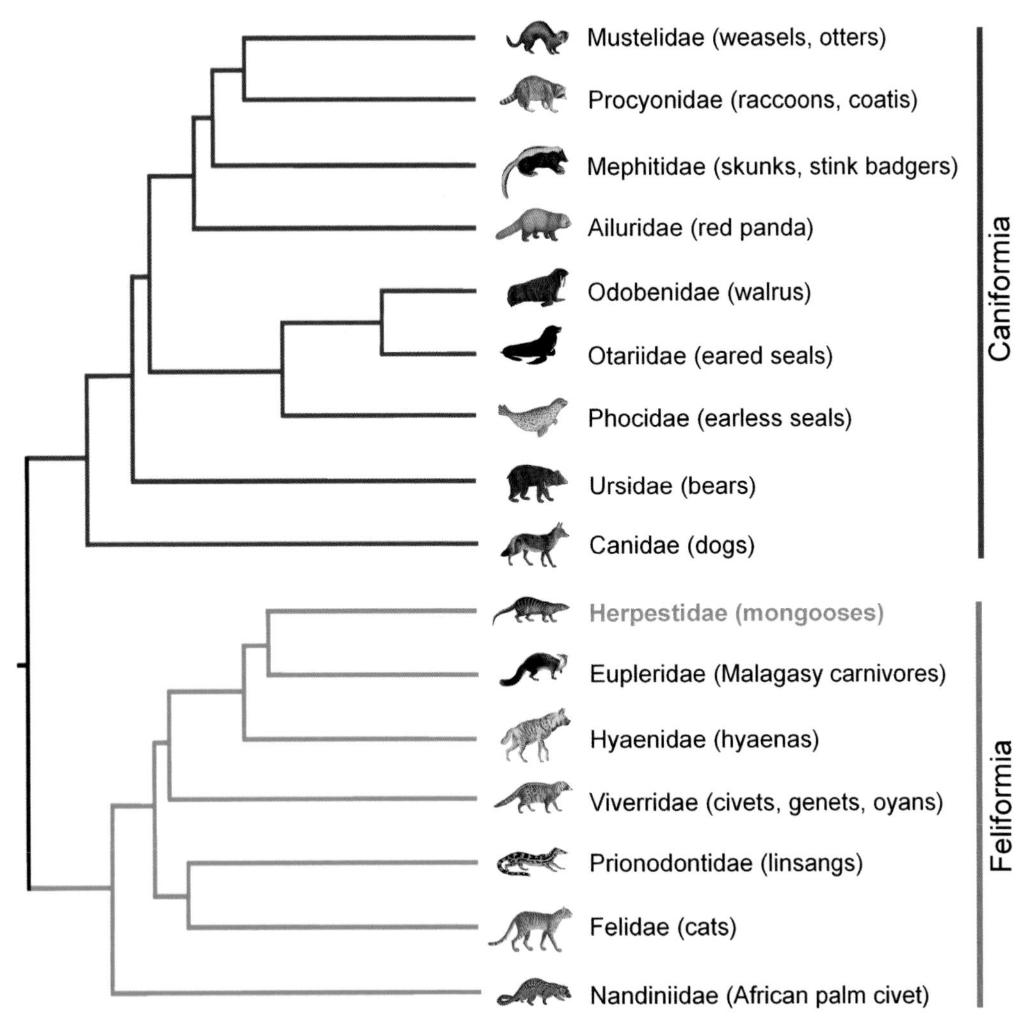

This phylogeny tree shows the relationships between the sixteen families of the order Carnivora, *which comprises two major groups: the* Caniformia *and the* Feliformia. *The mongooses (*Herpestidae*) belong to the* Feliformia *and their closest relatives are the Malagasy carnivores (*Eupleridae*) (© Andrew P. Jennings and Géraldine Veron; illustrations © Public Domain, Wikimedia Commons)*

within sixteen families that are assembled into two suborders: the Caniformia and Feliformia. Mongoose species are grouped together within the family Herpestidae, and placed within the suborder Feliformia, which also comprises the Malagasy carnivores, hyaenas, viverrids, linsangs, felids and the African palm civet.

The history of classifying mongooses is quite a long and complex one. In 1842, the French naturalist, René Lesson, placed the mongooses within a family that he called Ichneumonidae. The family Herpestidae, the group name that we use today, was actually established and described in 1845 by the French biologist Charles Lucien Bonaparte. In 1864, the British zoologist, John Edward Gray, further classified the Herpestidae into three subfamilies: Mungotinae, Herpestinae, and Galidiinae. This subdivision was supported by the British zoologist, Reginald Innes Pocock, in 1919, but he referred to the mongoose family as the Mungotidae.

Between 1845 and 1945, four naturalists, Étienne Geoffroy Saint-Hilaire, William Turner, William Henry Flower and Charles Torrey Simpson, in contrast to the classification schemes that we have just described, all placed the mongooses within a family called the Viverridae, which included civets and genets. This classification was followed by many other scientists, despite the anatomical work of William King Gregory and Milo Hellman, in 1939, which strongly suggested that the mongooses were not closely related to the other viverrids, and that they should be placed in a separate family, the Herpestidae, as earlier taxonomists had suggested. But the mongooses continued to be included in the Viverridae in many scientific publications and popular books, even after much scientific evidence supported the view that they should form their own family.

Over the last few decades, genetic studies have clearly revealed that the mongooses should be placed within the Herpestidae, as a separate family. In fact, genetic studies have also shown that the mongooses are actually more closely related to hyaenas (Hyaenidae) than to civets and genets (Viverridae).

There's just one more complication. Previously, several mongoose-like species, which are found only on Madagascar, were all grouped in the subfamily Galidiinae and placed with the other mongooses within the family Herpestidae (or sometimes in the Viverridae, when the mongooses were also included in this family). Genetic studies have now revealed that all the Malagasy carnivores (the mongoose-like forms and three other species, the cat-like fossa, the falanouc and the Malagasy civet, which were generally placed in the Viverridae) form a separate, albeit closely related, taxonomic group to the Herpestidae, called the family Eupleridae. It appears that around 20 million years ago, when no member of the Carnivora was present on Madagascar, the island was colonised by an ancestral African carnivore species, very closely related to mongooses, which then diversified into several mongoose-like, civet-like, and cat-like species. This is one clear example of what is known as convergent evolution, whereby some species can look very similar to those found in other parts of the world – although they are not, in actual fact, very closely related – simply because they have evolved and adapted to similar ecological and environmental conditions.

*The ringed-tailed vontsira (*Galidia elegans*) is a mongoose-like species that belongs to the Eupleridae, a family of carnivores endemic to Madagascar and the most closely related to the mongoose family (*Herpestidae*) (© Jason Woolgar)*

Within the mongoose family, Herpestidae, two subfamilies are currently recognised: the Mungotinae (social mongooses) and Herpestinae (solitary mongooses).

The subfamily Mungotinae consists of 11 African mongooses, which are all believed to be social, group-living species and share similar behavioural characteristics that might have been present in their common ancestor. Six social mongoose species primarily live in open savannah habitat, whereas the four cusimanses and the Liberian mongoose inhabit dense forests. The meerkat has a somewhat peculiar morphology and was previously recognised as a very distinct species – it was even placed in a separate subfamily called the Suricatinae – but the genetic evidence reveals that it is related to all the other social mongooses. The Liberian mongoose was only discovered by Western scientists in 1958, and its relationship to the other mongooses was initially unclear. It has recently been observed foraging in packs of 4–6 individuals, confirming that it is a social species, although the Liberian mongoose has smaller cheek teeth than other social mongooses, which may be due to the fact that it mainly eats earthworms.

The subfamily Herpestinae contains 23 mongoose species that live in Africa, the Middle East and Asia. This subfamily comprises solitary species, but a few field studies have revealed that some herpestine mongooses have more complex social systems than was

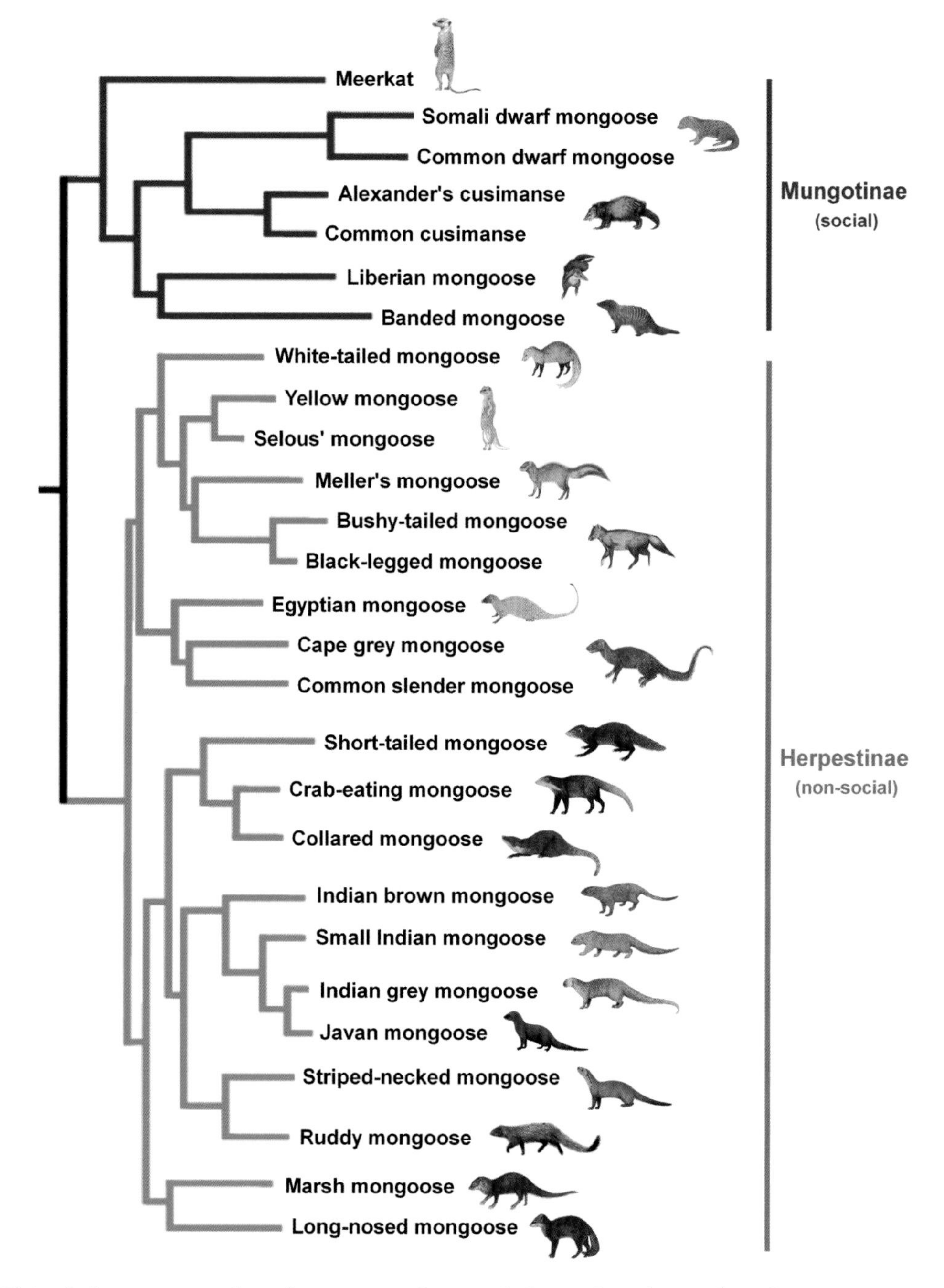

This phylogeny tree is based on genetic data and shows the relationships between 27 mongoose species, which are grouped into two subfamilies: Mungotinae (true social mongooses) and Herpestinae (solitary mongooses) (© Géraldine Veron and Andrew P. Jennings; illustrations modified from: Prater, S.H. (1971) The Book of Indian Animals. Bombay Natural History Society, Oxford University Press, Oxford; reproduced with the kind permission of the Bombay Natural History Society; Payne, J., Francis, C.M. & Phillipps, K. (1985) A Field Guide to the Mammals of Borneo. Sabah Society & WWF Malaysia, Sabah; Kingdon, J. (2015). Kingdon Field Guide to African Mammals. 2nd edition. Bloomsbury, London; reproduced with the kind permission of the artists)

previously thought. The yellow mongoose, for instance, displays some social behaviour, which prompted some scientists to include it within the subfamily Mungotinae, along with the other social mongooses. However, molecular studies have now confirmed that the yellow mongoose is related to the other solitary mongooses in the subfamily Herpestinae. Recent field studies have now revealed that although this species can be observed living in small groups, it does not exhibit all the social behaviour of the 'true' social mongooses, such as group foraging and baby-sitting. The poorly known Selous' mongoose was also originally considered a member of subfamily Mungotinae, but genetic studies have shown it to be closely related to the yellow mongoose, and so it too should be included in the subfamily Herpestinae.

The nine Asian Herpestinae mongooses were originally placed in the genus *Herpestes,* along with several African mongooses. Genetic studies have recently shown that the Asian mongooses all share a unique common ancestor, and are not closely related to the African *Herpestes* species. They should form a separate group, and have all been placed in the genus *Urva*. The small Indian mongoose and Javan mongoose were often considered as the same species, but genetic studies have now confirmed that they are separate species.

Interestingly, the mongoose species living on the island of Palawan, in the Philippines, was thought to be the short-tailed mongoose, based on physical characteristics, but recent genetic evidence has shown that, in fact, this species is a form of the collared mongoose. This is one example of where the physical appearance of an animal can sometimes be misleading when trying to classify it.

Within the subfamily Herpestinae, some mongoose species display a rare and interesting characteristic among mammals: the male has one less chromosome than the female. Normally, mammalian males have two sex chromosomes, known as X and Y. In some male mongoose species, however, the Y chromosome has attached itself onto a non-sex chromosome. This peculiarity has not yet been checked for all mongooses, and why and how this has happened is currently not known.

Taxonomic and systematic studies can often highlight the possible role of geographical and ecological barriers in shaping the past and current distribution of mongoose species. Over the past few millions of years, there have been dramatic climate changes that resulted in the sea level either rising or falling, which particularly affected some regions, such as Southeast Asia. When the sea level was lower than today, many landmasses, especially islands, were no longer separated from each other by bodies of water and mongoose species could then cross from one place to another. Climatic changes also caused some habitats, such as tropical forests, either to expand or contract, which obviously impacted those species that are dependent on them for their survival, and this has affected the distribution and speciation of mongooses in both Africa and Asia.

Despite the fact that recent genetic results have shed much light on the taxonomic status and evolutionary relationships of mongooses, not all species have been included in these molecular studies, and more are needed in order to clarify the complete picture. The number of mongoose species within the *Galerella* and *Bdeogale* genera, for instance, is

*[Left] The black mongoose (*Galerella sanguinea nigrata*) is a subspecies of the common slender mongoose that might be a separate mongoose species, but this requires further investigations (Namibia © Sfbergo CC BY-SA 3.0, Wikimedia Commons)*

*[Right] The bushy-tailed mongoose (*Bdeogale crassicauda*) comprises several subspecies, one of which, the Sokoke bushy-tailed mongoose, is sometimes considered a separate species, but further studies are needed to verify this (Tanzania © C. Fischer & Y. Hausser, HEPIA / ADAP)*

still uncertain. Unfortunately, several mongoose species are uncommon, elusive, or have been little studied in the wild, which has made it very difficult to obtain genetic samples from these species. While DNA can also be obtained from museum specimens, this is often degraded and not as good as that obtained from fresh material.

Defining species and subspecies is more than a fascinating academic endeavour: it has very important implications for conservation purposes, as it is crucial to know what biological entity one is trying to protect and conserve. For instance, small isolated populations are particularly threatened with extinction, and ascertaining whether these are taxonomically unique might be a crucial factor when deciding whether any money should be targeted towards conservation efforts to save these populations.

EVOLUTION OF MONGOOSES

The fossil record of mongooses is unfortunately quite poor, making it very difficult to construct their evolutionary history, and unearthing this story is made even harder by the difficulty of identifying old fossils as belonging to mongooses.

The oldest fossil remains that were recognised as an ancient mongoose species is a creature called *Leptoplesictis,* which lived in Africa during the early part of an evolutionary time period called the Miocene, which extends from 23 to 5 million years ago. This animal then later occurred in Europe during the middle Miocene. Only fragments of teeth and jaw have been discovered, so it is not possible to do a full reconstruction of this animal. Several *Leptoplesictis* species have been described – some were small, about the size of a small Indian mongoose that lives today, and others were larger.

The current fossil evidence suggests that during the Miocene, mongooses migrated from Africa to Asia – unambiguous mongoose fossils have been found in late Miocene deposits in both Africa and Pakistan – and exchanges between these two regions could have persisted until the early part of the Pliocene, a period that extends from 5 to 2.5 million years ago. Soon afterwards, the Arabian Peninsula became largely desert and the Strait of Gibraltar opened up, which no doubt severely impeded any possible migration of mongooses between Africa and Eurasia.

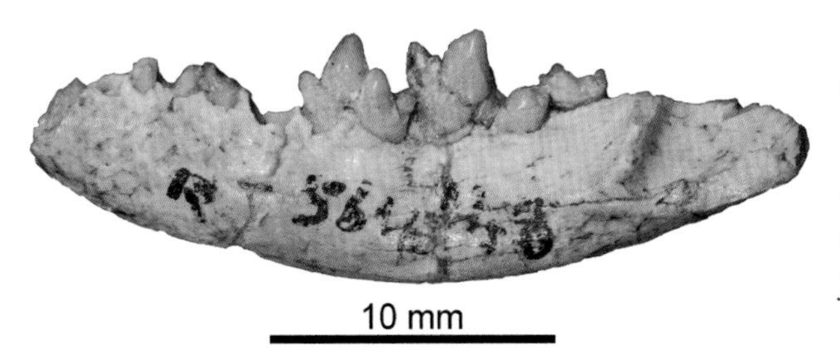

This is part of the left lower jaw of Leptoplesictis rangwai *from Rusinga Island, Kenya. It is around 17.9 million years old and is one of the oldest mongoose fossils to be discovered (© Lars Werdelin/National Museums of Kenya, KNM-RU 15990B)*

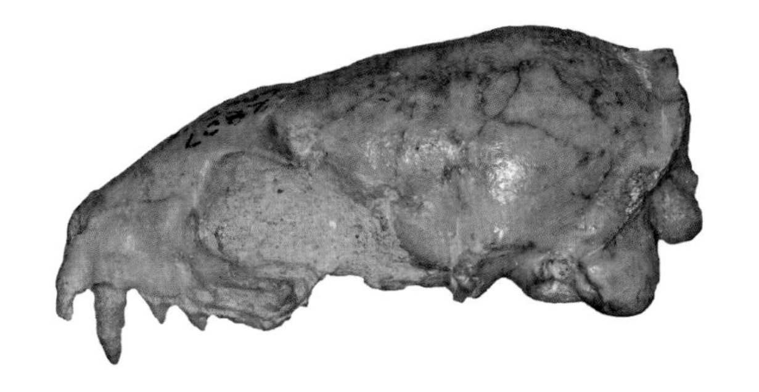

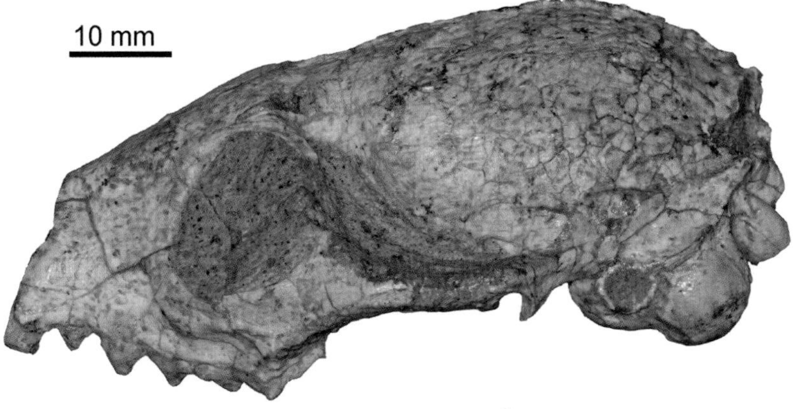

Two mongoose fossil skulls from Laetoli, Tanzania, both around 3.7 million years old. Top: Helogale palaeogracilis. *Bottom:* Herpestes palaeoserengetensis *(© Lars Werdelin/National Museums of Kenya, LAET 75-2807A and LAET 78-5435A)*

During the last few million years, climate changes caused some forested areas either to expand or contract, or become a more open savannah-like habitat – a mosaic of trees and grassland that does not have a completely closed canopy like a dense woodland. The proliferation of fossil genera that have been discovered in Africa from 5 to 0.8 million years ago suggests that these habitat changes triggered a diversification of African mongoose species.

Environmental changes, particularly the Ice Ages that began 2.6 million years ago, also affected the diversification and distribution of Asian mongoose species, especially across Southeast Asia. During glacial periods, the climate was more seasonal, which drastically altered the distribution of forested and more open habitats. The sea level was also lower, resulting in land bridges being formed between different landmasses, including present-day Sumatra, Borneo, Java and numerous smaller islands, as well as the mainland. However, despite various present-day islands being connected during glacial periods, major rivers flowed between the land areas of Java, Borneo and Peninsular Malaysia, which could have restricted the movement of mongooses across this region. All these environmental changes facilitated the movement of some species from one area to another, whereas for others, ecological and physical barriers impeded dispersals and isolated some populations. Although genetic studies on present-day species can give some hint as to what happened in the past, the fossil record is very scarce in this region, so it is currently not possible to use physical evidence to track precisely what impact these environmental changes had on the dispersal and speciation of mongoose species.

MORPHOLOGICAL AND ANATOMICAL CHARACTERISTICS

Mongooses generally have a pointed face, a long slender body, short legs and a bushy tail, and range in weight from the 200g common dwarf mongoose to the 5kg white-tailed mongoose. They all share certain anatomical features, such as the structure of their ear bones and an anal pouch, which is a pocket around the anal opening that contains scent glands.

The colour and pattern of an animal's coat can help camouflage it from predators and, not surprisingly, mongooses living in open, sandy habitats are usually light-coloured and forest-dwelling species are generally dark. Mongooses have a fur coat with no spots (and rarely stripes) but instead, their pelage has a uniform colour which, depending on the species, varies from light-grey or yellow to brown or black, with legs or feet that are sometimes darker than the rest of the body. The coat hairs are usually long and coarse, and each hair is often ringed with different colours that give the fur a grizzled appearance. The banded mongoose and meerkat do have stripes or streaks across their back, whereas others, including the stripe-necked mongoose and crab-eating mongoose, have a distinctive neck stripe, which could act either as a device to warn potential predators that it can expel an obnoxious secretion from its anal glands, or as an indicator of a 'safe' area on the body, to which young mongooses or amorous males can direct their bites during play-fighting or courtship.

[Above] The yellow mongoose is a medium-sized mongoose with a pointed head, small ears, quite short limbs, and a long bushy tail (South Africa © Emmanuel Do Linh San)

[Below] The common slender mongoose has a long weasel-like body (South Africa © Emmanuel Do Linh San)

[Above] The common dwarf mongoose has a domed head and is the smallest mongoose in the world (Zimbabwe © Emmanuel Do Linh San)

[Below] The common cusimanse is a small mongoose that has a shaggy coat, short legs and a distinctive long snout (© Chris Stuart and Mathilde Stuart)

[Above] Yellow mongooses live in habitats that often have a sandy substrate and yellowish-coloured vegetation. This photograph shows how well they can blend in with their background, which makes it more difficult for predators to spot them. These two individuals differ slightly from each other in colouring, illustrating the variation that can exist within a mongoose species (© Luke Hunter)

[Below] Mongooses generally have a uniformly coloured coat, with no stripes or spots. The crab-eating mongoose, however, has a white stripe on each side of its neck (India © Vijay Anand Ismavel)

The stripe-necked mongoose has a black neck stripe. A few other mongoose species have neck stripes, and their function is still unclear (India © Kalyan Varma)

The banded mongoose is one of two mongoose species that have dark bands across its back (the other is the meerkat) (Tanzania © C. Fischer & Y. Hausser, HEPIA / ADAP)

A mongoose's muzzle is fairly long, with well-developed *vibrissae* (whiskers) around the mouth, above the eyes, on the cheeks and below the chin. Whiskers are thick hairs that have numerous nerve endings at the base of the hair follicle, making them very touch-sensitive and helping an animal to navigate or locate food in very dark situations, when it is too difficult to see. The Indian grey mongoose is known to have sebaceous glands at the base of the vibrissae that produce a honey-like substance, which might possibly be used to scent mark objects. This feature may be present in other mongooses, as some have been seen rubbing their cheeks on various supports. The nose is quite large in some mongooses, but it is not really known if their sense of smell is very well developed. Like cats, meerkats have been observed making a grimace known as 'flehmen', which is provoked by strong-smelling substances. This action helps bring odours to a pouch in the roof of the mouth that is lined with receptors, known as the Jacobson's organ.

The eyes of mongooses have horizontally elongated pupils. This likely extends the visual field in the horizontal plane, which would be particularly beneficial to open-habitat species that scan their surroundings for predators and food. Some diurnal mongooses appear to have good colour vision that helps them recognise certain food items and predators in daylight. Nocturnal animals commonly have a reflective layer on the outside of the retina called the *tapetum lucidum*, which reflects light back into the eye so that it can be detected again by the photoreceptors. This reflecting layer enables nocturnal carnivore species to see in very low light conditions, and is responsible for the eyeshine that is seen when a bright light is directed into their eyes – anyone driving at night has probably seen the shining eyes of an animal caught in the beam

The head of this meerkat shows its pointed muzzle and the tufts of whiskers around its mouth, which are thick hairs with sensitive nerve endings (South Africa © Emmanuel Do Linh San)

[Left] The eye of this yellow mongoose has a horizontally shaped pupil, which is typical of the mongooses (South Africa © Emmanuel Do Linh San)

[Right] The white-tailed mongoose is a nocturnal species and has a reflective layer at the back of its eye called the tapetum lucidum, *which helps it see in the dark. This feature is found in other mongooses, and is responsible for the bright eyeshine which is seen when a light is directed towards the animal (© Kalyan Varma)*

of their headlights. Mongooses that are nocturnal, such as the marsh mongoose and white-tailed mongoose, have very obvious eyeshine, whereas this is much less striking in the mainly diurnal species, such as the common cusimanse. In fact, some diurnal mongooses species, such as the yellow mongoose, are thought not to have a *tapetum lucidum*.

Mongoose ears are small and rounded, and lack a *bursa*, which is a folded purse of skin on the outer rim that is found in other carnivores, such as cats. However, the yellow mongoose has large ears with a small depression that may represent a vestigial bursa. The ears of some mongooses that burrow, such as the meerkat, can be closed by using the ear ridges.

The feet of a mongoose are long and narrow, with generally five digits on each foot, although a few species have reduced or missing *pollex* (thumb) and *hallux* (big toe). All mongooses, except for the marsh mongoose and the short-tailed mongoose, have partly-webbed feet, but the webbing between the toes is often much reduced and difficult to see. The unwebbed fingers of the marsh mongoose are long and sensitive, and are used to probe for food in cloudy water or soft mud. The footpads of mongooses are not very distinct and are mostly naked or covered with a few hairs. Unlike a cat, the long, curved claws of mongooses are not retractable, and are used for digging burrows and for rooting out food items embedded in the ground or within rotting logs.

[Above] As in all mongooses, the ear of this meerkat does not have any folded skin that forms a small pocket, which is seen in cats. Meerkats have internal ridges that can close their ears when digging (South Africa © Emmanuel Do Linh San)

[Below] The forefeet of an Egyptian mongoose have strong claws which enable it to dig out invertebrate prey (South Africa © Jarryd Streicher)

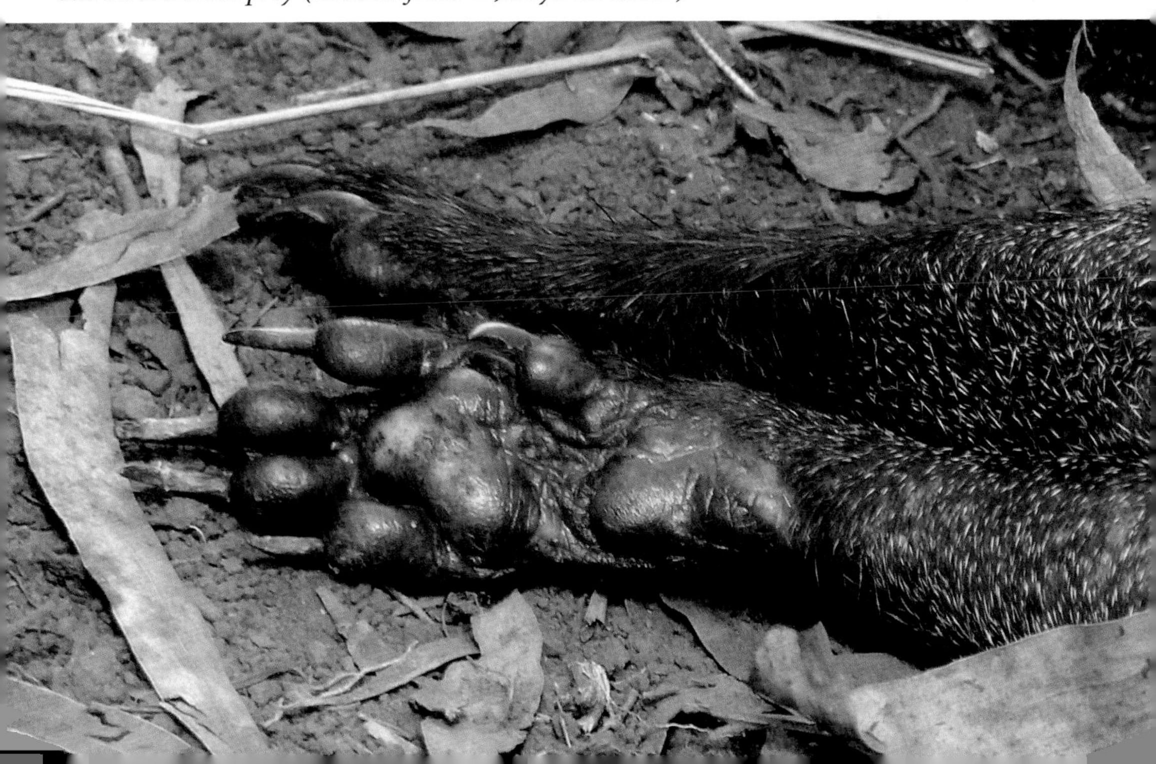

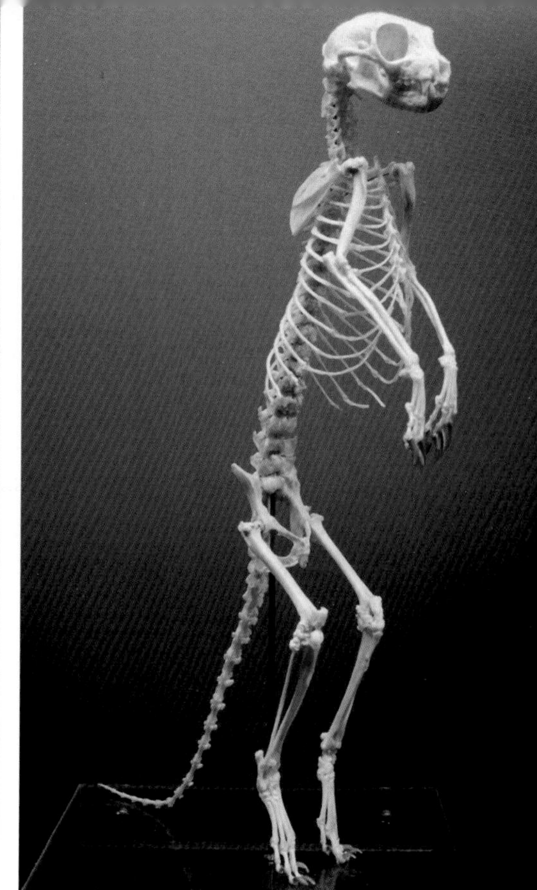

[Left] The long claws of a meerkat are used for digging (South Africa © Emmanuel Do Linh San)

[Right] The skeleton of a meerkat in the standing position, showing how it uses its tail for added support (© Chris Dodds CC BY-SA 2.0, Wikimedia Commons)

As the animal's name suggests, the long bushy tail of a white-tailed mongoose is white, but some individuals have a black tail (Zambia © MC Schaeffer CC BY-SA 3.0, Wikimedia Commons)

A mongoose's tail can be as long as the body, as in the Indian grey mongoose, or is particularly short, such as in the short-tailed mongoose, where it is less than half its body length. The tail is particularly bushy in some of the larger mongooses, such as the bushy-tailed mongoose, and, in the white-tailed mongoose, the tail differs in colour from the body.

To be an efficient predator, mongooses must be agile enough to be able to run and dig, and also have the capability to make a large variety of movements for capturing, killing and eating live prey. The consequence of having this great versatility is that mongooses generally have a quite unspecialised skeleton, unlike other mammal species, such as gibbons, that live in the tops of trees and have long forelimbs that are specifically adapted for swinging from branch to branch. Mongooses, however, spend most of their time scurrying around on the ground, and often forage or rest in confined spaces, such as burrows, so they tend to have quite short limbs. Mongooses are digitigrade, which means that they stand and move on their toes, similar to cats and dogs, rather than on the whole sole of their feet, like a bear.

The skull of most mongooses is elongated, with a fairly short muzzle. In the meerkat, the skull is especially round and short, and the eye orbits are placed in a forward-looking position, which gives them good binocular vision for judging distances – meerkats detect predators by standing on their hind legs and moving their head from side to side. The bony case surrounding the middle ear, called the auditory bulla, has two chambers, separated in feliform carnivores by a bony wall. In some mongooses, such as the yellow mongoose, the anterior chamber is inflated and larger than the posterior chamber, and also has a characteristic 'C' shape. The anterior chamber is also well inflated in the other mongooses, but remains smaller than the posterior chamber.

Mongooses generally have 40 teeth with, on each upper and lower jaw, six incisors, two canines, eight premolars and four molars. In several mongoose species, however, there are fewer premolars, and the total number of teeth can vary from 36 to 38. The small incisors are mainly used for nipping food items, and the canines are used to grasp and

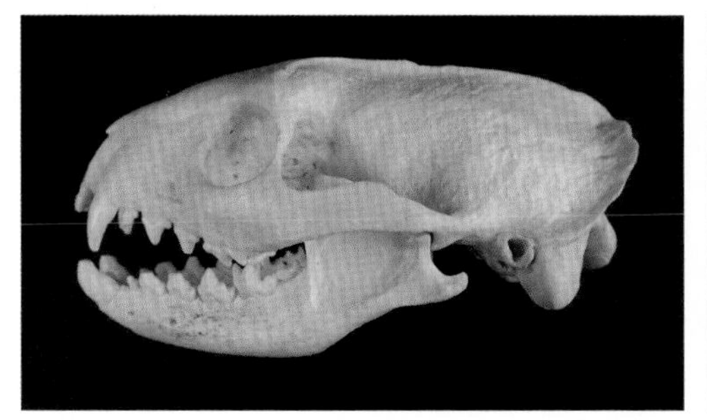

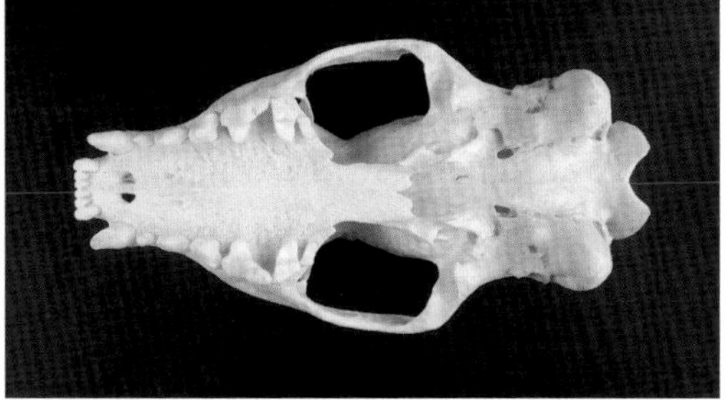

[Left] The skull of an Egyptian mongoose showing its elongated shape, large canines and well-developed carnassial teeth (© Géraldine Veron /MNHN-ZM-MO-1986-456)

[Right] The ventral view of an Egyptian mongoose skull showing the large inflated bone cases (tympanic bullae*) that are located at the back of the skull, at the right of the picture (© Géraldine Veron /MNHN-ZM-MO-1986-456)*

kill prey by piercing the neck or skull. As we have already mentioned, the characteristic carnassial teeth in carnivores are specially adapted to cutting through flesh with a scissor-like action, and this is the case for most mongoose species. The premolars help hold prey or, like the molars, are used to crush food items.

The teeth of each mongoose species are somewhat modified and adapted to what they eat. In those species that feed on small mammals, for instance, such as the yellow mongoose, the canines are long, the carnassials are well developed and the cusps of the premolars and molars are sharp. The meerkat feeds largely on insects, for which it has sharp teeth, but the carnassial shear is very poorly developed. Meller's mongoose and the bushy-tailed mongoose also have an insectivorous diet, and their cheek teeth have low, rounded cusps that are used to crush and grind their hard-bodied prey. In the marsh mongoose, the cheek teeth are very robust and are used to break the shells of crabs.

Mongooses, like many other mammal species, have anal glands that secrete a strong, foul-smelling liquid. These glands are paired sacs that are located under the skin of the anal region and which open into the rectum through a short canal. In mongooses, the skin around the anus is folded inwards to form an anal pouch, which can be everted to apply the anal gland secretions to various objects. These secretions contain characteristic chemicals that not only help convey the identity of an animal, and possibly its general health and stress level, but are sometimes used in defence against predators – skunks are well known for being able to spray the contents of their anal glands at intruders. Although mongooses cannot spray their secretions, some species, such as the marsh mongoose, may eject a strong-smelling brown fluid from their anal glands when stressed.

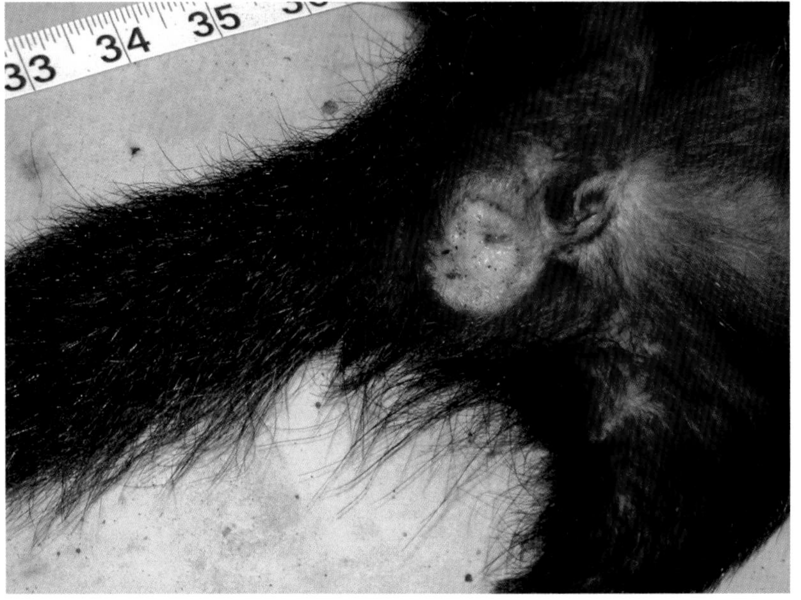

The anal poach of mongooses is a pocket of folded skin around the anus opening, into which flow the smelly secretions from the anal glands (short-tailed mongoose, Peninsular Malaysia © Géraldine Veron and Andrew P. Jennings)

Male mongooses have a bony feature called a *baculum*, a small bone found within the penis. Males of many other carnivore species have a penis bone that helps in penetrating a female's vagina and providing vaginal stimulation. The shape of the *baculum* varies among the mongoose species, each showing a distinctive form.

2. The lifestyle and behaviour of mongooses

Where do mongooses live?

Mongooses occupy a broad range of habitats, from open areas, such as semi-desert, savannah and grassland, to dense temperate and tropical forest. Some mongoose species are found in specific habitats (e.g. the short-tailed mongoose mainly occurs in rainforest) while others, such as the small Indian mongoose and Cape grey mongoose, can live in many different habitats. Within broad habitat types, such as forest or savannah, mongooses can be found within a diverse range of ecosystems; for instance, the meerkat and Somali dwarf mongoose occur in dry arid areas, whereas the marsh mongoose and crab-eating mongoose prefer the banks of rivers and streams, as well as swamps and wetlands.

[Left] The banded mongoose mainly lives in open habitats such as savannah and grasslands (Kenya © Emmanuel Do Linh San)

[Right] The short-tailed mongoose lives in dense rainforests (Malaysia © Jayasilan Mohd-Azlan)

Although mongooses can be found over a wide altitudinal range, from sea level to the tops of mountains, some species seem only to occur within certain elevations; the short-tailed mongoose, for instance, mainly lives in lowland areas. Jackson's mongoose is one of the highest recorded species, at 3,300 metres.

Finding food is one factor why a particular mongoose species lives in a specific habitat. For instance, rivers and streams are associated with aquatic prey, such as frogs and fish, whereas open habitats, such as savannah and bushy areas, support high densities of insects and small mammals. What you eat, then, largely dictates where you live, or vice versa.

How do several mongoose species manage to live together in the same geographical area, when there is strong competition for food? This is achieved through an ecological principle called 'niche partitioning' – different mongoose species occupy separate ecological niches, often by preferring a different habitat or diet. The Cape grey mongoose and yellow mongoose both occur in South Africa, but the Cape grey mongoose lives in bush habitat, where it feeds on small mammals, and the yellow mongoose prefers open habitat, where its insect prey is abundant. Field studies in South Africa have also found that the slender mongoose selects forested areas, the Egyptian mongoose prefers open habitat, the marsh mongoose is closely associated with water, and the banded mongoose lives in scrub forest. Mongooses can also avoid each other, and other competitors, by having different periods of activity: some species are nocturnal, while others are diurnal.

Mongooses also live alongside other carnivore species, with which they may compete for food or are in danger of being preyed upon. The presence of other carnivore species, then, might well affect where a mongoose lives. In the Serengeti, there appears to be fewer white-tailed mongooses in areas where there are high numbers of foxes and aardwolves. Mongoose species living in open habitats are very visible to predators, especially those that have keen sight, such as birds of prey. In fact, the high risk of being attacked by a

A few mongoose species, such as this stripe-necked mongoose, will look for food in human refuse sites (India © Kalyan Varma)

predator in open areas is thought to be a major selective pressure favouring the evolution of sociality amongst some mongoose species. However, the forest-living cusimanses and the Liberian mongoose are also social species, so other factors may account for their social behaviour, as we will explain in the next section.

Mongooses can have a significant impact on the habitat in which they live. Those species that dig burrows or root in the ground to find food can alter the physical structure of their environment, which can then have consequences for other wildlife. The Liberian mongoose disturbs the soil while foraging for earthworms, and this could affect the germination of different plants seeds.

Most mongoose species are rarely seen near human settlements, whereas others, such as the small Indian mongoose and banded mongoose, can live close to humans and may feed on garbage as a supplemental food source. Some mongooses, including the white-tailed mongoose, may use buildings, woodpiles, and other human structures as resting sites or places to raise young.

Social organisation

Do mongooses form social groups, like a pack of wolves, or do they live largely on their own, like the solitary cats? In fact, many mongoose species are solitary, others are social (or semi-social), and some solitary species can sometimes form social groups (depending on the local circumstances), which highlights the extraordinary diversity of social

[Left] The common slender mongoose is a solitary species that lives in Africa (© Laila Bahaa-el-din)

[Right] The crab-eating mongoose is a solitary species that lives in Asia (© Vijay Anand Ismavel)

The meerkat is a true social mongoose, with groups of up to 50 individuals living together. Each pack member helps raise young and watch out for predators (South Africa © Emmanuel Do Linh San)

organisation that exists within this carnivore family. However, we do not yet know the precise details about the social system of every mongoose species, as many have not been well studied in the wild.

The 11 mongoose species within the subfamily Mungotinae are all thought to be social, group-living species. Group sizes commonly range from 3 to 30, and comprise adult males and females, and young. Social mongooses cooperate to raise their young, including babysitting, feeding and grooming pups, and have a coordinated vigilance system for detecting predators, with individuals taking turns to stand guard.

Why do some mongoose species live in groups, whereas others do not? Cooperative hunting is a common explanation for why large carnivores, such as wolves and lions, form social groups: several individuals, acting together, can bring down prey larger than they could kill on their own, which can then feed the whole group, including any dependent young. But mongooses rarely work together as a team to capture prey that is larger than themselves. Instead, it seems that the availability of food is one prime driving force for mongooses being social. Many solitary mongoose species include a large proportion of small vertebrates (small mammals, reptiles, amphibians, and birds) in their diet, and usually, there are too few prey within their foraging area to support a group of mongooses living together; in other words, there is just not enough of this type of food to go around to feed several mouths at once. Also, one mongoose foraging alone is less likely to scare away a mouse or a lizard than a large group thrashing through the undergrowth. In contrast, invertebrates, such as insects, can be very abundant in some habitats; they are also much less sensitive to disturbance by several individuals foraging together, and invertebrates can quickly replenish an area after they have been harvested. So any mongoose species that can exploit this situation can form social groups: there is enough food to go around, and having another individual beside you does not completely ruin your chances of eating a meal. Interestingly, the yellow mongoose is a species that usually forages alone and eats both insects and small vertebrates, and can exhibit some social, group-living behaviour.

Predation risk is likely another instrumental factor for sociality in mongoose species, particularly for those that live in open savannah habitat. Mongooses are quite small animals that are vulnerable to large predators, such as larger carnivores, snakes and raptors, and a mongoose moving around in open habitat is particularly at high risk since it is much more visible to preying eyes. Being in a group can then be beneficial through coordinated vigilance for an approaching predator. In the meerkat and common dwarf mongoose, group members take turns to go 'on guard' by standing on an elevated termite mound or tree to watch for predators. This individual gives a specific call, the 'watchman's song', to inform other group members which mongoose is on guard, and upon sighting a predator, it will then give a specific alarm call. This guarding helps reduce mortality within the group, since several dedicated eyes are scanning the horizon at all times. This vigilance system demonstrates the high levels of cooperation and communication that have evolved in social mongoose species.

[Left] A meerkat searching the sky for aerial predators, such as eagles and hawks (South Africa © Emmanuel Do Linh San)

[Right] To look out for dangerous enemies from a high vantage point, common dwarf mongooses will sometimes climb a tree (Botswana © Laurent Vallotton)

Common dwarf mongooses will perch on the tops of termite mounds to observe their surroundings (South Africa © Julie Kern)

A team of mongooses may also be able to repel some predators. When confronted with a large snake, groups of meerkats or banded mongooses will bunch together and engage in to-and-fro behaviour, and this 'super-organism' will often scare off the snake. Several banded mongooses have also been seen leaping up to a meter in the air in an attempt to bite airborne marabou storks that have grabbed a pup, and groups of common dwarf mongoose have successfully rescued members that had been captured by an aerial predator.

The four cusimanse species and the Liberian mongoose are also within the subfamily Mungotinae, but they live in forests, unlike the other social mongooses that live in open habitats. These five species have been seen in groups, but we do not know if their social behaviour is as sophisticated as the other well-studied social species. Not much is known about their diet either, except that it appears that earthworms and other invertebrates are major components. European badgers live in groups in places where they eat mostly earthworms, so the abundance of earthworms within foraging areas could be a major factor for allowing these five mongooses to be social. Without any knowledge of the predation risks that these species face in their forested environment, we do not yet know if they gain any anti-predator benefits of being in groups through increased vigilance. Another possibility for why these five forest mongooses are social is that, before they moved into dense forests, their ancestors lived together in groups in more open areas, and so their current social behaviour could just be a legacy of their past evolutionary history.

[Above] The Liberian mongoose is a social, group-living species that lives in dense forests (© Kevin Schafer)

[Upper right] The Somali dwarf mongoose is a poorly known social species that lives in eastern Africa (© ZooParc de Beauval)

[Right] The banded mongoose is a social species that is found in open habitats across Africa (© Laila Bahaa-el-din)

The 23 species within the subfamily Herpestinae are all thought to live mainly solitary lives, with males and females defending separate territories, and those of males commonly overlapping one or two females. In solitary mongooses, males and females only come together for a short time to mate – a few days at most – and the only extended period of cohabitation is between a mother and her offspring, which can last a few months. Recent field studies, however, have revealed that some solitary mongoose species can have more complex social interactions than were previously thought. The Egyptian mongoose and small Indian mongoose can form groups when there are

The yellow mongoose is a species that usually forages alone, but often lives in groups for the rest of the time. However, it is not closely related to the true social species, and individuals do not adopt specific roles, such as babysitting or guarding against predators. (South Africa © Emmanuel Do Linh San)

very abundant food resources, such as human garbage dumps. The yellow mongoose exhibits some semi-social behaviour: although they tend to forage alone during the day, several individuals will sleep together at night in the same den, and some groups may cooperate to raise their young. But yellow mongooses do not take turns to stand guard for predators or babysit pups in the den, so there is less individual role specialisation than in the true social mongooses.

MOVEMENTS AND ACTIVITY

Mongooses are mainly terrestrial, but a few species, such as the banded mongoose and cusimanses, may occasionally climb trees, and many mongooses, including the marsh mongoose and crab-eating mongoose, can swim well – but there are no truly tree dwelling or aquatic species in the mongoose family.

The activities of mongooses are mainly concerned with finding food, a place to rest or sleep, and a mate during the breeding season; they also spend time grooming themselves, patrolling their home ranges or territories, and interacting with other mongooses, either other group members (in the social species), potential mates, or young pups.

[Left] Social mongooses, such as the common dwarf mongoose, will spend time grooming other pack members (Zimbabwe © Emmanuel Do Linh San)

[Right] A common slender mongoose deeply asleep on the branch of a tree. Finding a safe place for resting is crucial for a mongoose, as they are small animals that are vulnerable to a wide variety of predators (South Africa ©Emmanuel Do Linh San)

Meerkats will sometimes spend time fighting each other as a means of maintaining a dominance hierarchy (South Africa © Emmanuel Do Linh San)

Activity period and social organisation of mongooses

Species	Scientific name	Activity	Social organisation
Marsh mongoose	*Atilax paludinosus*	Nocturnal	Solitary
Bushy-tailed mongoose	*Bdeogale crassicauda*	Nocturnal	Solitary
Jackson's mongoose	*Bdeogale jacksoni*	Nocturnal	Solitary
Black-legged mongoose	*Bdeogale nigripes*	Nocturnal	Solitary
Alexander's cusimanse	*Crossarchus alexandri*	Diurnal	Social
Ansorge's cusimanse	*Crossarchus ansorgei*	Diurnal	Social
Common cusimanse	*Crossarchus obscurus*	Diurnal	Social
Flat-headed cusimanse	*Crossarchus platycephalus*	Diurnal	Social
Yellow mongoose	*Cynictis penicillata*	Diurnal	Semi-social
Pousargues' mongoose	*Dologale dybowskii*	Diurnal	Social
Kaokoveld slender mongoose	*Galerella flavescens*	Diurnal	Solitary
Somali slender mongoose	*Galerella ochracea*	Diurnal	Solitary
Cape grey mongoose	*Galerella pulverulenta*	Diurnal	Solitary
Common slender mongoose	*Galerella sanguinea*	Diurnal	Solitary
Somali dwarf mongoose	*Helogale hirtula*	Diurnal	Social
Common dwarf mongoose	*Helogale parvula*	Diurnal	Social
Egyptian mongoose	*Herpestes ichneumon*	Diurnal	Solitary
White-tailed mongoose	*Ichneumia albicauda*	Nocturnal	Solitary
Liberian mongoose	*Liberiictis kuhni*	Diurnal	Social
Gambian mongoose	*Mungos gambianus*	Diurnal	Social
Banded mongoose	*Mungos mungo*	Diurnal	Social
Selous' mongoose	*Paracynictis selousi*	Nocturnal	Solitary
Meller's mongoose	*Rhynchogale melleri*	Nocturnal	Solitary
Meerkat	*Suricata suricatta*	Diurnal	Social
Small Indian mongoose	*Urva auropunctata*	Diurnal	Solitary
Short-tailed mongoose	*Urva brachyura*	Diurnal	Solitary
Indian grey mongoose	*Urva edwardsii*	Diurnal	Solitary
Indian brown mongoose	*Urva fusca*	Nocturnal	Solitary
Javan mongoose	*Urva javanica*	Diurnal	Solitary
Collared mongoose	*Urva semitorquata*	Diurnal	Solitary
Ruddy mongoose	*Urva smithii*	Diurnal	Solitary
Crab-eating mongoose	*Urva urva*	Diurnal	Solitary
Stripe-necked mongoose	*Urva vitticollis*	Diurnal	Solitary
Long-nosed mongoose	*Xenogale naso*	Diurnal	Solitary

The majority of mongooses are active during the day, and few species are nocturnal, which again highlights the behavioural diversity of this family. Mongooses are capable of moving rapidly and covering great distances within a day, and can probably cross their territory or home range within a 24-hour period. White-tailed mongooses can move up to 4 kilometres per hour, and over the course of a day, meerkats travel up to 6 kilometres, and banded mongooses cover 10 kilometres or more.

Mongooses may travel quite large distances to search for food and other resources, such as den sites or a mate, and banded mongooses can cover over 10 km each day (© Kalyan Varma)

The time period during which a mongoose is awake and active most likely depends on what it eats. Mongooses feed mainly on live animals, which are much easier to detect by sight and sound when they are moving, so it would make good sense for a mongoose to be active at the same time as its prey. For instance, most small mammals, such as mice, scurry around in the undergrowth at night, and a mongoose that catches small mammals to eat would have a much better chance of seeing and hearing them if it were also nocturnal. Most insect species are more easily spotted crawling along the ground or on plants during the daytime, and since the meerkat, common dwarf mongoose and banded mongoose are three species that eat mainly insects, it is very likely that their diet largely explains why they are active during the day. Unfortunately, we do not know enough about the diet and activity of most mongooses to determine whether these two are closely linked for all species. Other factors, such as avoiding competitors or certain predators, may also play an important role in determining a good time to be active. Many large carnivore species are active at night, and if being killed by them poses a high risk for a mongoose, it would need be active only during the day to stay steer clear of these predators.

When mongooses are not searching for food or seeking a mate, they rest or sleep, and finding a safe place in which to hide is very important for a mongoose, since it is very vulnerable to larger predators while it is not alert. Several mongoose species are known to use burrows, which they may dig themselves, whereas others, such as the short-tailed mongoose, may crawl inside a hollow log. Mongooses also need secure den sites for when they are rearing young, and in some habitats, suitable rest or den sites may sometimes be in short supply. Common dwarf mongooses use termite mounds for sleeping at night; and they may also use them for rearing young during their first month of life, or as look-out posts and refuges from predators. These mounds are so important to this species that a higher number of common dwarf mongooses are found in areas where there are more termite mounds.

Termite mounds are very important places for common dwarf mongooses, which use them as a shelter for sleeping in, as a safe place for raising young, and as a vantage point for looking out for predators (South Africa © Julie Kern)

[Left] Common dwarf mongooses may sometimes rest inside a hollow tree (© Laila Bahaa-el-din)

[Right] Yellow mongooses den in communal burrows, which they either dig themselves or take over from other animals (Botswana © Claude Fischer)

Home ranges and territories

The area within which a solitary mongoose or a group of social mongooses lives is known as a home range, or a territory, if this area is actively defended. Many factors can affect the size of a home range, including the size and sex of an animal, habitat quality, diet, food availability and group size (in social mongooses), and so home range sizes vary both within and between species. For the few mongoose species that have been studied in the field, recorded home range or territory sizes are usually less than a few square kilometres. The riparian marsh mongoose has home ranges that tend to be long and thin, following a stream or river.

Social mongooses actively defend territories, which generally do not overlap. When two rival groups meet, the smaller one will generally retreat without any physical aggression, although skirmishes can sometimes occur and some individuals may be killed during such encounters. If two similar-sized banded mongoose groups meet, members of both packs will bunch together and approach each other with caution, stopping frequently to stand upright and stare at their opponents. Individuals will then rush forward and fan out to chase an opponent or engage in a one-to-one fight. Confrontations are often violent, involving repeated bites and scratches with the front paws, until one of the combatants bolts. If one group becomes scattered, it will sometimes retreat, then bunch together and advance again – in this way, fights between evenly-matched groups can sometimes last for over an hour.

When two meerkat groups meet, rival pack members may fight each other. These confrontations can sometimes be very violent, even ending in the death of some individuals (South Africa © Emmanuel Do Linh San)

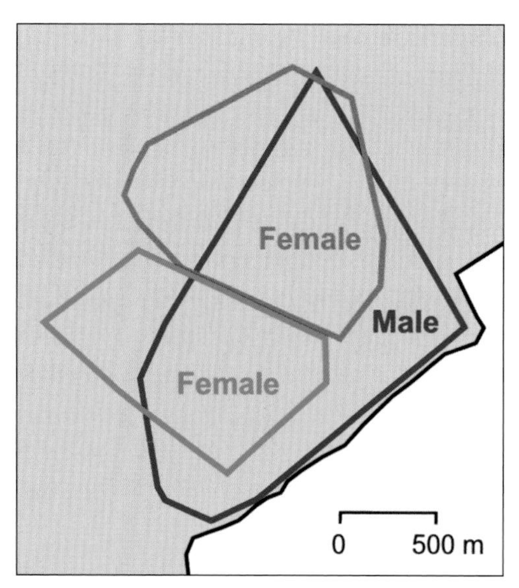

This map shows the home ranges of three short-tailed mongooses that were radio-tracked in a rainforest in Peninsular Malaysia (grey area is forest, white area is farmland). This is a solitary mongoose species; the home range of a male overlaps at least two females, which have separate home ranges (© Andrew P. Jennings and Géraldine Veron)

In solitary mongooses, male home ranges may overlap those of one or more females, and so are often substantially larger than female home ranges. In some mongoose species, the home ranges of each sex can overlap considerably those of the same sex. The home ranges of male small Indian mongooses, for instance, can overlap by as much as 84%, and those of females by 37%; in contrast, the home ranges of female white-tailed mongooses can overlap considerably, whereas male ranges are fairly exclusive.

Mongooses mark their home range or territory by using faeces, urine and the smelly secretions from their anal glands, which are often deposited in places where rival individuals or groups will most likely encounter them.

Habitats with higher food densities generally support smaller home ranges or larger group sizes. For example, the home ranges of the white-tailed mongoose in the Serengeti (Tanzania), where there is a high density of beetles, are smaller than those in Bale (Ethiopia), where beetle density is lower.

The day-to-day use of a home range is strongly affected by food availability. Mongooses usually forage in different areas on subsequent days, and this allows the replenishment of food in those places already harvested. In exceptional circumstances, when there is far more food available than one mongoose can eat, it is not worth the time or effort to defend this abundant food resource, and some solitary mongooses may temporarily allow intruders to enter their home range to take advantage of a bonanza-like situation: three common slender mongooses were observed spending several days together feeding on the large mass of fly larvae that were crawling over a giraffe carcass, and up to seven Kaokoveld slender mongooses were seen feeding upon the swarm of flies on the remains of a greater kudu. Several Egyptian mongooses and small Indian mongooses may feed together on garbage dumps.

Diet and foraging behaviour

Mongooses eat mostly live animals, and only a few species will also consume a small amount of plant material, particularly fruits and berries. Since a solitary mongoose is too small to tackle large prey (except for the amazing ability to kill snakes larger than themselves), their diet is usually restricted to small vertebrates (small mammals, birds, reptiles, amphibians and fish) and invertebrates (including insects, spiders, slugs, crabs, snails and millipedes), and this diet can vary considerably within and between species.

[Left] Yellow mongooses predominantly eat insects, preferring termites, beetles, locusts and grasshoppers. However, they are opportunistic feeders and will also hunt other invertebrates, as well as rodents, birds, reptiles and amphibians (South Africa © Emmanuel Do Linh San)

[Right] Meerkats mainly eat invertebrates, such as beetles and termites, but occasionally they will feed on small vertebrates, including rodents and frogs (Namibia © Schnobby CC BY-SA 3.0, Wikimedia Commons)

The skull morphology of a mongoose species reflects its specific diet. Since killing and eating vertebrates requires more effort than feeding on invertebrates, mongooses that have a more vertebrate-based diet tend to have a more robust skull than insectivorous species. The type of diet is also reflected in the shape of the different teeth. The marsh mongoose has particularly robust premolars for crushing crustaceans, such as crabs, whereas in the Liberian mongoose, which specialises in eating earthworms, the teeth are much smaller. Mongooses that mainly eat small vertebrates have particularly long, strong canines for delivering a killing bite, and also a well-developed carnassial shear for slicing flesh.

Where a mongoose lives and looks for food obviously influences what it can eat. The mainly insectivorous yellow mongoose forages in open short-grass plains, where invertebrate prey is abundant, whereas the Cape grey mongoose lives in bush habitat, where small rodent prey is plentiful. Marsh mongooses live in swamps and marshes, and forage along streams and river edges, where they feed on crabs, frogs, toads, and fish. In drier areas, or during drier spells, they switch to more terrestrial prey, such as mammals and birds, and their diet also differs between coastal and inland areas.

Common dwarf mongooses forage on the ground in search of small prey, especially insects and other invertebrates (South Africa © Julie Kern)

Mongooses often adjust their diet to feed on what is most available. Egyptian mongooses are opportunistic predators and will feed on the most abundant prey. The diet of the small Indian mongoose can vary from principally insects in a desert region, to crabs in mangrove forests, to rats in canefields. Prey availability may also affect the timing of reproduction. In most species, the birth of young coincides with the rainy season, which is likely to support high prey densities.

A mongoose's diet can also change with seasons. The white-tailed mongoose switches from termites and ants in the dry season to dung beetles in the wet season. The diet of the crab-eating mongoose is dominated by insects in summer and autumn, crustaceans in winter, and reptiles in late spring and early summer. The common slender mongoose eats predominantly insects from May to September, and small vertebrates from October to April.

As predators of live animals, mongooses need to use all their senses to locate and capture prey, but whether they rely mostly on sight, hearing, smell or even touch, largely depends upon what they are trying to catch and where – to find and grab insects in open savannah during the day, capture rodents in dense undergrowth at night, or snatch small fish from a river, all require different senses to be used.

Mongooses are active feeders, constantly moving around. They hunt for prey mostly on the ground, but some may also search in burrows, or within the beds of streams and rivers, and a few species may occasionally forage in small trees. Solitary mongooses will trot along and then pounce upon, or chase, any small vertebrate or invertebrate prey that they flush out from the undergrowth, or they may stop frequently to pick invertebrates off vegetation. A group of social, insectivorous mongooses will spread out to forage, and each individual will then scratch at the surface and periodically stop to dig. The cusimanses root around in forest litter with their long snouts, and the marsh mongoose uses its dexterous forefeet to probe for prey in mud or shallow water.

[Left] Meerkats will dig intensively for invertebrate prey, sometimes even disappearing from view in the deep hole that they have dug (South Africa © Emmanuel Do Linh San)

[Right] The ruddy mongoose may use its sense of smell to search for prey. Very little is known about its diet, but they have been seen feeding on mice, snakes, and birds (India © Kalyan Varma)

To kill small vertebrates, mongooses tend to bite the skull, but some species may use a neck bite on larger prey. Mongooses usually kill and eat invertebrates starting at the head. Some mongooses have a characteristic method of dealing with hard prey, such as beetles, crabs, molluscs, and birds' eggs. A mongoose will pick up the object in its forepaws, orient itself so that its rear is facing a rock or other hard surface, and then throw the item between its rear limbs to crack or smash the object, making the contents accessible for eating. The marsh mongoose and crab-eating mongoose will also throw hard objects down to the ground vertically from a standing position, rather than horizontally through the hind legs.

Mongooses are perhaps most famed for their ability to kill snakes, immortalised in traditional fables about the small Indian and Indian grey mongooses. The speed and agility of the mongoose is said to be instrumental in their ability to avoid being bitten and to overcome a snake, although some mongoose species are highly resistant to snake neurotoxins. However, the remains of snakes are rarely found in the scats or stomach of mongooses, which suggests that snakes are not a major food item. Other dangerous or toxic prey types eaten by mongooses include scorpions, centipedes, and millipedes. The small Indian mongoose is also known to eat the toxic parotid glands of toads.

[Left] Some mongooses will occasionally eat snakes, and they have become renowned for their ability to fight snakes much larger than themselves (© Public Domain; Brooks, Elbridge S. 1901, Animals in action; studies and stories of beasts, birds and reptiles; their habits, their homes and their peculiarities. *Boston, Lothrop Pub. Co)*

[Right] Many mongooses will eat eggs, which they throw backwards between their rear limbs to crack open on a hard surface. Some species may smash eggs by throwing them downwards from a standing position (© Public Domain; Illustrated London Reading Book, *1851)*

Cooperative hunting by a pack of wolves or lions enables them to capture large prey, but this does not usually occur in the social mongooses as they eat mainly invertebrate animals, which do not require coordinated hunting to capture; instead, individuals respond aggressively to approaches by others when they are feeding or digging a foraging hole. However, groups of common cusimanses and common dwarf mongooses may occasionally work together to hunt a large snake or rat, which may then be shared among themselves. Among the solitary mongoose species, two small Indian mongooses – possibly a mother teaching her young – have been observed hunting crabs together, with one turning over stones and the other attacking the crab, while several Egyptian mongooses were once seen excavating rabbits from their dens, although it is not clear if this hunting was coordinated.

Scavenging is not uncommon in the mongoose family; for instance, the stripe-necked mongoose will scavenge on sambar and Indian hares, and the Indian grey mongoose has also been observed feeding on carrion. The occurrence of large mammal material in the scats and stomach of some other species also suggests scavenging, as mongooses do not often capture animals much larger than themselves. Scats of the long-nosed mongoose and marsh mongoose sometimes contain ungulate remains, while porcupines have been found in Cape grey mongoose scats, and these are unlikely to have been killed by those mongooses. Occasionally, some species – the small Indian mongoose, for instance – may prey on domestic livestock, such as fowl, and this can bring them into conflict with people.

Some mongoose species will occasionally scavenge for food, such as this common slender mongoose feeding on a carcass (© Nadine E. Cronk)

How do mongooses communicate?

For mongooses to communicate with each other, using either visual displays, sound or scent, they need good vision, hearing and sense of smell. Although we do not know a great deal about mongoose morphology and physiology, they are thought to have good senses.

Visual communication, such as behavioural displays, can only happen when individuals are able to see each other at close range. In some social mongooses, for instance, an individual displays its dominance by standing erect on its hind legs, while another acknowledges its inferior status by assuming a low, crouching posture. When two groups of meerkats confront each other, they adopt a 'war dance,' in which each individual arches their back and stiffens their legs, erects its tail, and rocks back and forth on their front and back legs. When attempting to repel a terrestrial predator, a pack will bunch together and move to-and-fro, thus appearing as a large, fearsome 'super-organism'.

A tail is a useful visual signalling device. It can reveal the emotional state of an individual or its dominance status – the twitching tail of an angry cat, or a cowering dog holding its tail between its hind legs, are familiar to pet owners. The white-tailed mongoose has a tail that is usually coloured differently from the rest of the body, which would certainly help emphasise any tail signals – it arches its tail when scent marking and when close to other mongooses. The yellow mongoose has a conspicuous white tail tip and uses different tail positions to indicate a range of emotions, such as 'satisfaction' (the tail is held straight tail, with a downward kink in the middle), and 'uncertainty' (a straight tail, with an upward kink in the middle).

The white-tipped tail of the yellow mongoose is used as a visual display towards other group members, and the way it is held indicates the emotional state of an individual (South Africa © Emmanuel Do Linh San)

Information on mongoose vocalisations is generally scant since most mongooses are solitary and rarely seen in the wild. Nevertheless, field and captive studies have yielded some information on the vocalisations produced by various species, particularly social mongooses, which tend to have a wider vocal repertoire and make a greater use of vocal communication than do solitary species. However, most mongooses at the very least have distinct vocalisations for aggression, alarm, and pain.

Twelve distinct vocalisations have been identified for the small Indian mongoose, including a squawk and a growl. The white-tailed mongoose mutters while digging for insects, barks if threatened, and screams when hurt. The vocal repertoire of marsh mongooses includes bark-growl threats, excitement bleats, and moan/bleats that may fulfil a contact role. The Egyptian mongoose has seven distinct vocalisations that include a deep, sharp growl that causes other individuals to flee, and a bark/spit given during mating or fighting. Yellow mongooses will emit an alarm bark to elicit others to run to a bolthole, and a high-pitched scream during fights.

Vocalisations are better known for the intensely studied meerkat, banded mongoose, and common dwarf mongoose. These species all give a regular contact call while foraging, which serves to maintain group cohesion. These contact calls enable each individual to remain with the group without constantly having to look around, and may also identify its neighbour. Dominant females will emit a 'moving out' call to elicit the rest of the group to follow. Growls and spits are used when another group member attempts to steal food. Shrill 'war' cries will alert group members to a rival group, and elicit bunching and charges. Short, sharp alarm calls are given upon sighting a potential predator, which will trigger rapid evasive behaviour in all the group members. 'Worry' calls warns the group of dangers that might warrant increased vigilance. Some vocalisations are often accompanied by characteristic body behaviours – banded mongooses will both spit and lunge, and growl with a hunched body shape.

[Left] Social mongooses, such as the common dwarf mongoose, have a wide range of vocalizations that they use to communicate with each other (South Africa © Julie Kern)

[Right] Vocalizations are an important means of communication between group members in the semi-social yellow mongoose (South Africa © Emmanuel Do Linh San)

Young mongooses also vocalise. Pups give a continuous 'begging' call while they follow foraging adults, and an intense, high-pitched 'give-me-food' call when a nearby adult finds a food item. It is likely that similar begging occurs in the solitary mongoose species. In addition to begging calls, pups can also give simple warning calls from an early age.

The alarm calls of both the meerkat and common dwarf mongoose vary depending on the identity of the predator, and whether it is approaching on the ground or in the air; they can also indicate the level of danger. These varying alarm calls determine how a group responds. When a meerkat gives a mammalian predator call, the group will run to the nearest bolthole and then look around. In response to an avian predator call, meerkats will crouch and freeze, and look to the sky. A snake alarm call will elict individuals to come together and mob the snake. If a danger is perceived as being very close, the signaller gives a 'high urgency' variation of the specific alarm call.

In contrast to canids and felids, mongooses do not appear to use vocalisations to defend their territory or to attract mates, probably due to their smaller size and greater vulnerability to predation. Instead, scent is used for these purposes. For a family whose ancestry probably lies in a solitary, nocturnal lifestyle, it is not surprising that smells play a major role in communication. Scents have the advantage of being persistent, lasting days to weeks, and they work in the dark. Scent signals can be deposited in the environment either by using secretions from the anal scent glands or through different chemicals within faeces and urine. Different smells can reveal the identity of an individual, give clues of its health and sexual status, attract other mongooses for mating, and even be used to repel predators.

Mongooses possess an anal pouch that can be turned inside out to deposit secretions from the anal scent glands. These glands produce odoriferous carboxylic acids, which are long-chained in the Egyptian mongoose and short-chained in the small Indian mongoose. In the Egyptian mongoose, male secretions include a major component that is not present in those of females. In contrast, the relative concentrations of acids differ among all small Indian mongooses, and these different odours possibly allow individuals to identify each other. Marsh mongooses respond differently to anal gland secretions (and also scats), depending on the sex of the individual that left the scent. Different carboxylic acids evaporate at different rates, so the chemical signal changes over time – common dwarf mongooses can differentiate between anal gland secretions based upon the age of the secretion. An anal scent mark, therefore, can provide information not only on the sex and identity of the marker, but also on the timing of its deposition. The gland secretion of common dwarf mongooses can be detected for up three weeks, which is about how long it takes a group to complete a circuit of its territory, suggesting that this secretion is used in marking territorial boundaries.

Anal scent marking is commonly used to mark prominent objects within a home range, or to mark other individuals within a group. It is usually achieved by horizontally dragging the anus along the ground or an object, but depending on the actual structure of the anal glands in each species, the scent can also be deposited on vertical objects using a characteristic reverse handstand or cocked leg. In some social mongooses, dominant individuals commonly scent mark more often than subordinates. The Egyptian mongoose, crab-eating mongoose and marsh mongoose can forcefully eject the anal gland secretion under stress.

Cheek marking has been seen in some mongoose species, including the yellow mongoose and marsh mongoose. In the common dwarf mongoose, cheek gland secretions persist for only two days and do not contain any personal identity markers, but do elicit a hostile response from other individuals.

Mongooses also mark their territory using urine and faeces, which not only contain characteristic smells produced by the anal glands, but also act as a visible marker. Faeces are often deposited together in specific latrine sites, and in some species, these latrines are concentrated in places where it is more likely that they will be discovered by other mongooses. Meerkat groups often share one latrine site with their neighbours, which probably allows them to monitor the status of each other. In the banded mongoose, groups respond to the scent of faeces and urine from another group by 'worry' calling and sniffing, and over-marking the site by defecating, urinating or anal marking. The communication role of scent marking is also apparent in the white-tailed mongoose, which deposits scats at latrines that may be used by several neighbouring adults, and will frequently urinate and anal mark when there is another mongoose nearby. Marking likely also plays a role in sexual advertisement. Faeces and urine contain hormones, particularly oestrogen in females, which can inform others of an animal's reproductive status.

[Left] Mongooses commonly use their anal gland secretions and urine to scent mark prominent objects within a home range or territory (yellow mongoose, South Africa, © Emmanuel Do Linh San)

[Right] Mongooses have scent glands on their cheeks that they can use to mark objects or other individuals (yellow mongoose, South Africa © Emmanuel Do Linh San)

Breeding

The mongoose family incorporates a variety of breeding systems and care of young. In general, male mongooses may mate with multiple females during their fertile period, and females often mate with more than one male. Females are receptive (in oestrus) for a few days, and can have several fertile periods throughout the year (polyoestrus). Induced ovulation takes place, in which the release of the ova (eggs) is triggered by mating. These traits likely originated from the solitary lifestyle of ancestral mongooses, in which lone individuals searched for mates over a wide area during the breeding season. Under these circumstances, regular receptive cycles, a relatively long receptive period, and ovulation triggered by copulation, would all improve the chances of successful matings and fertilisations.

In solitary mongoose species, adult males and females live alone. It is likely that a female advertises her receptivity through hormonal cues in her urine (and possibly in other secretions), which are monitored by a male as he traverses his home range, which overlaps that of at least one female. Finding a mate is obviously much easier for social mongooses, and oestrus periods often involve fights between competing individuals within the group. In meerkat and common dwarf mongoose groups, there is a dominance hierarchy, and usually only the dominant male and female breed; however, a few subordinate females may also mate and give birth to litters, but their young are often killed by the dominant pair. In contrast, almost all the members of a banded mongoose group breed, and several females will give birth in the same den, on the same day. In these three species, both males and females may temporarily leave their own territory in order to mate with neighbouring individuals. In the semi-social yellow mongoose, females within a group give birth approximately 4–10 days apart.

[Top left] Indian brown mongooses are a solitary species, but a male and female will stay together for several hours, or a few days, during the period when a female is receptive (India © Divya Mudappa)

[Top right] The normally solitary stripe-necked mongoose will spend a short time with their mate during the reproductive season (India © Kalyan Varma)

[Lower left] Unlike some other social species, most members of a banded mongoose group will mate and produce pups (© Laila Bahaa-el-din)

[Lower right] Almost nothing is known about the reproductive behaviour of the Gambian mongoose (© Ondřej Machač)

During the period when a female is receptive (oestrus), a male will remain in close association for hours or days. Copulation itself may involve chasing, playing, or fighting, prior to the male mounting the female, which usually involves the male clasping the female with his forelimbs forward of the female's pelvis and thrusting his pelvic region; he may also grasp a female's neck in his mouth. Copulation may last from a few seconds to a few minutes, and since this may not always be successful, a male is likely to copulate a number of times with a female while she is in oestrus. During this period, a female may also mate with other males.

A pair of yellow mongooses mating (Rotterdam Zoo © Magalhães/Public Domain)

Mongoose females typically produce one or more litters per year (up to five in the banded mongoose), and births often occur when there is plenty of food available. Depending on the species, the pregnancy period or gestation can vary from 6–11 weeks. Female banded mongooses are receptive again within one week of giving birth, and in meerkats and yellow mongooses, it seems that oestrus occurs again while the female is still suckling a litter.

Female mongooses have four or six teats, and the maximum litter size is usually six. Young mongooses are called pups and are born in a secure den site, such as in a burrow, or within a termite mound. Solitary mongooses tend to have small litters of 1–3 pups that remain in the birth den for approximately ten weeks. In contrast, social mongoose females often have larger litter sizes, and the pups emerge from the den at around four weeks.

Mongoose pups are born blind and deaf, and sparsely furred. Their weight at birth can range from as little as 20g in the banded mongoose, up to 125g in the marsh mongoose. A pup's eyes open at around two weeks, and if the pups need to be moved to another den, they are usually picked up by the scruff of their neck. As far as we know, only the mother raises the young in solitary species, whereas in social mongooses, other females and males, called helpers or allocarers, will assist in caring for the pups. The larger litter sizes and reduced period of dependence in the social mongooses may be due to the additional care and food that is provided by these helpers.

[Left] Pups of the Indian grey mongoose (India ©Vinc3PaulS CC BY-SA 3.0, Wikimedia Commons)

[Right] Pups of the yellow mongoose (South Africa © Emmanuel Do Linh San)

[Upper left] Young common slender mongooses (South Africa © Emmanuel Do Linh San)

[Upper right] Adults and pups of the common dwarf mongoose (South Africa © Julie Kern)

[Left] A young common dwarf mongoose (South Africa © Julie Kern)

These young meerkats can already adopt a standing position to observe their surroundings (South Africa © Emmanuel Do Linh San)

In the social mongooses, hormones and pheromones likely play an important role in the control and synchronisation of births. Dominant individuals probably use both aggression and hormones to try and suppress the breeding of subordinates. In the meerkat and common dwarf mongoose, the dominant breeders make the lowest contribution to caring for the young, whereas all adult banded mongooses provide care to the communal litter. Non-breeding helpers will babysit and groom the pups, and some females may even nurse young that are not their own by producing milk and allowing them to suckle.

Why in social mongooses do non-breeders care for young that are not their own? Helpers are often related to one of the breeding pair, which means that they will share some of their genes with the pups. So by caring for someone else's young, a helper is actually ensuring that a few of its genes are passed on to future generations.

Young mongooses are usually suckled in the den, where they remain until they are ready to emerge. In the social mongooses, a group normally leaves at least one babysitter with the pups while the rest of the pack goes away to forage. However, suckling females rarely babysit, as they need to forage to fulfil the heavy energetic demands of providing milk. Unlike larger social carnivores, such as canids, mongooses do not bring back food to the young, except in the yellow mongoose, where large prey items, such as rodents, are delivered to the den once the pups have been weaned.

In some yellow mongoose groups, pups are looked after by their parents and other individuals (South Africa © Emmanuel Do Linh San)

[Left] Young mongooses are dependent on adults until they are weaned and can forage for themselves (yellow mongoose, South Africa © Emmanuel Do Linh San)

[Right] In meerkat groups, adults will take turns to babysit young at the den, while the rest of the pack goes away to forage. The role of babysitters is to guard pups and keep them warm (Namibia © Claude Fischer)

When pups are old enough to go on foraging trips, any food that is captured on the spot is shared with them. In most species, this is done only by the mother, but in the social mongooses, other adult males and females (breeders and non-breeders) will provision pups with food items. Finding and catching live prey is a difficult task and learning the skills for this takes young mongooses several months to master. In social mongooses, group members teach pups the foraging skills that they need. At first, an adult will kill and give the food to the pups, either whole or in pieces, but gradually, over time, young mongooses learn each step of this process themselves, and as the pups gain the skills required to forage independently they are fed less. Unfortunately, we do not currently have any information on how solitary mongooses raise their young after they have been weaned.

Play is the major way for a young animal to learn important skills that will be used as an adult. However, in meerkats, it seems that play does not increase social cohesion, nor reduce aggression between individuals, so at least in this species, play behaviour may not have much impact on the development of social skills.

Social bonds are very important between young yellow mongooses and their parents (South Africa © Emmanuel Do Linh San)

A mother and young yellow mongoose (South Africa © Emmanuel Do Linh San)

Young meerkats will beg for food from adults (South Africa © Emmanuel Do Linh San)

[Left] Play is an important activity for learning social behaviours (yellow mongoose, South Africa © Emmanuel Do Linh San)

[Right] Two pups of the common dwarf mongoose playing (South Africa © Emmanuel Do Linh San)

Fighting and playing are important for the development of social behaviours in meerkats (South Africa © Emmanuel Do Linh San)

In the meerkat and banded mongoose, the young are dependent upon group members until they are about three or four months old, after which they are old enough to find food themselves. A solitary mongoose mother will stay with her pups for a longer time, around six months. In the majority of species, young mongooses reach adult size and attain sexual maturity between nine months and two years.

As soon as a young mongoose is independent, it may disperse from its natal area in order to reduce the possibility of inbreeding with close relatives. In the solitary mongooses, both male and female young disperse, although often a young female will remain close to or within her mother's territory. Among the social mongooses, it is usually young males that disperse to join other groups, although females may also voluntarily leave or be aggressively evicted by a dominant female.

Predators and relationships with other animal species

Being relatively small, mongooses are vulnerable to many predators, especially snakes, raptors and other carnivores. For instance, predators of banded mongooses include African rock pythons and martial eagles, while marabou storks and monitor lizards have been seen taking pups. Meerkat predators include black-backed jackals, tawny eagles, pale chanting goshawks and snakes. A Cape grey mongoose was seen being killed by a caracal (a medium-sized cat). As the smallest of the mongooses, common dwarf mongooses are potential prey to a great variety of predators, including jackals, steppe monitor lizards and brown snake eagles.

African mongooses, including this yellow mongoose, are vulnerable to a wide range of predators, including the black-backed jackal (© Nadine E. Cronk)

[Left] Meerkats will watch for predators and signal any danger to the rest of the pack (South Africa © Emmanuel Do Linh San)

[Right] Like the meerkat, the yellow mongoose will adopt a standing position to look out for danger (South Africa © Emmanuel Do Linh San)

Although the mortality of young can be particularly high, mongooses have the potential to be relatively long lived under favourable conditions. In general, mongooses can live to around 8–12 years in the wild, and up to 20 years in captivity.

The social mongooses display an impressive level of cooperation with each other and they also sometimes cooperate or interact with other species. Liberian mongooses are often found in close association with monkeys, such as sooty mangabeys, and will flee in response to their anti-predator warning calls. Banded mongooses groom warthogs to feed on the ticks and other insects infesting their skin, and will respond to the alarm calls of numerous other species, especially plovers. Meerkats respond to the alarm calls of fork-tailed drongos and yellow-billed hornbills. Common dwarf mongooses will respond to the alarms calls that hornbills give out when a raptor approaches. This is a fascinating relationship that benefits both parties. Early in the morning, hornbills will perch upon the termite mound in which the group of mongooses has taken refuge overnight, waiting for them to rouse. Before long, the mongooses start to emerge and begin searching for food – in fact, they will delay their departure if there are no hornbills around. As the group of mongooses forage through the shrubs and grass, the hornbills perch on branches above, snapping up any insects that are disturbed. If a hornbill spots an approaching raptor, it will then utter an alarm call, sending the mongooses scurrying to the nearest shelter. In this mutualistic association, the hornbills get an easy meal, and in return, the mongooses benefit from the greater vigilance that the hornbill's higher vantage point gives them. Should the mongooses happen to sight a raptor first and give out an alarm call, the hornbill will also flee – so each species benefits from the vigilance of the other. Hornbills will even utter an alarm call for raptor species that do not prey on them but are predators of the mongooses. Common dwarf mongooses are also known to form a similar relationship with other co-foraging bird species, such as fork-tailed drongos.

A group of yellow mongooses may sometimes share a burrow with meerkats or ground squirrels. Yellow mongooses and meerkats that live together will cooperate in looking out for predators, which increases their chances of detecting dangerous animals. Meerkats may also share a burrow with other species, including springhares, Cape grey mongooses and common slender mongooses, apparently without any hostilities.

Yellow mongooses may share burrows with ground squirrels (South Africa © Claude Fischer)

3. MONGOOSES AND HUMANS

MONGOOSES IN HUMAN CULTURE, MYTHOLOGY AND FOLKLORE

Mongooses have had a long history of association with people, particularly within the region stretching from northern Africa to India. This is not too surprising, as some mongooses are often found in close proximity to humans, exploiting the abundant food in garbage dumps or preying on the mice, rats and cockroaches that live around people's homes. With their perceived characteristics of courage, cunning, strength, ferocity and curiosity, mongooses have played a significant role in human culture since early times and are featured in many myths, legends, and stories.

Mongooses were well known to the ancient Greeks and Romans, and were venerated in Egypt, where they were considered a sacred animal, from at least 2500 BC. Ancient Egyptians believed that mongooses break crocodile eggs and that without the mongoose the number of crocodiles would be so great that no-one would be able to approach the Nile. The sun-god Re once transformed himself into a mongoose to fight Apophis, the serpent of the netherworld. The mongoose god in the mortuary temple of Amenemhat III (1860–1814 BC) represented the spirits of the netherworld. In Letopolis, the mongoose god was equated with the falcon-god Horus and in Heliopolis with the creator-god Atum. Mongooses were often embalmed in large numbers, and their mummies have been found inside small bronze statues of the lion-headed goddess Sekhmet. Many Egyptian mongoose mummies were discovered at Tanis, in the Nile delta of Egypt. Often these mummies were packed in small coffins, the lids of which were decorated with either paintings or models of the living animal.

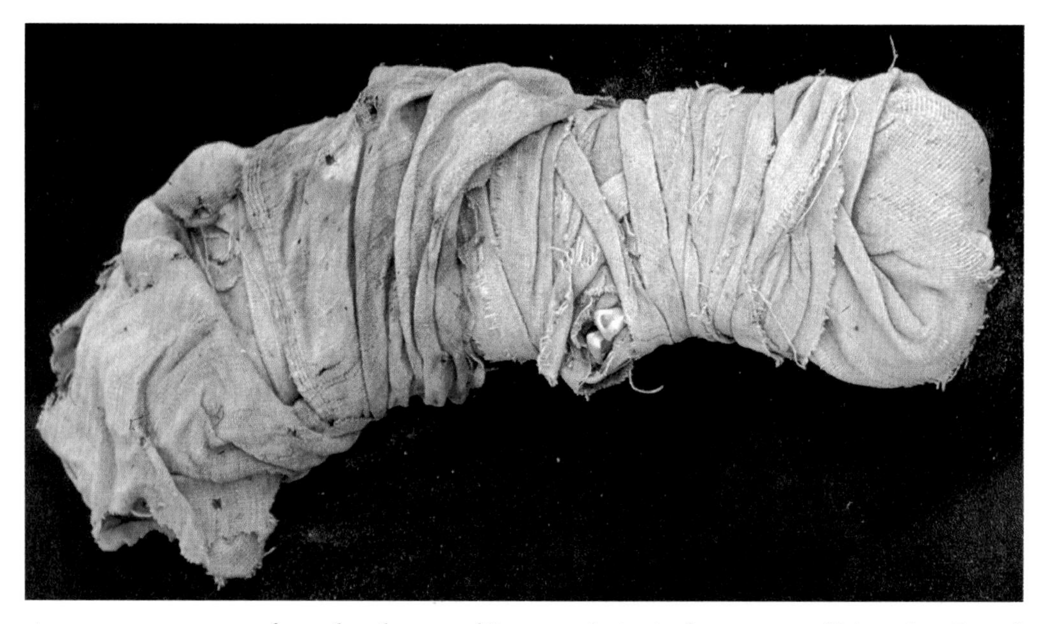

A mongoose mummy from the plateau of Saqqara (principal cemetery of Memphis, Egypt) and dated to the Ptolemaic and Roman period (305 BC–476 CE) (© Cécile Callou)

Mongooses have been the subject of artworks and have appeared as bronze figures, on coins, or as figurines of human beings with mongoose-like heads. Representations of the Egyptian mongoose or ichneumon can be found on the walls of tombs and temples of Thebes and Saqqara in Egypt; the earliest date from the Old Kingdom (2686–2181 BC). Mongooses were often shown raiding birds' nests for their eggs or clambering through papyrus thickets, on the prowl for snakes, small rodents and invertebrates. Two Old Kingdom scenes show mongooses being held by their tails as men lower them into the waters of the Nile, suggesting that tame mongooses could have been used to flush out birds from papyrus thickets during the hunt. The greatest number of mongoose representations that were left by the ancient Egyptians date from the Greco-Roman period (after 332 BC), when the cult of sacred animals was at its height. These include small bronzes that may have been produced in large numbers. Some of these are of the mongoose walking, with the legs in such a position that all four can be seen from the side. Others show the animal in the sitting position with the sun disc and the *uraeus*, or sacred asp, on its head.

A wall painting in an Egyptian tomb showing a hunter with a mongoose and a dog (Chapel of Baqet I, no. 29, Beni Hasan, Egypt © Linda Evans, Australian Centre for Egyptology, Macquarie University, Sydney; reproduced with permission from the Ministry of State for Antiquities, Egypt)

[Left] A bronze standing Egyptian mongoose from the Ptolemaic era (305–30 BC) (© Rama CC BY-SA 2.0 fr, Wikimedia Commons)

[Centre] A bronze Egyptian mongoose with forepaws raised in adoration, dated to around 500 BC (© Walters Art Museum/Public Domain)

[Right] A bronze standing Egyptian mongoose, with uraeus and solar disc, from the Ptolemaic era (305–30 BC) (© Rama CC BY-SA 2.0 fr, Wikimedia Commons)

Mongooses have featured prominently in Middle and Far Eastern religions, frequently as guardians of wealth. In Indian Hindu mythology, the mongoose is regarded as the natural foe of the *nagas* (serpents) and a guardian of the jewels and treasures lying under the earth. Mongooses were conceived of having wrested the wealth from the possession of the serpents – their stomachs were considered a good repository for hidden treasures and the riches a suitable tribute to Kubera, the son of a sage, who was given immortality by Brahma and made a god of wealth and guardian of all the treasures of the earth. In the first century BC, Kubera was sculpted with a mongoose-shaped purse (probably made from mongoose skin) in his left hand, and is later depicted holding a mongoose in his hands. In Buddhist mythology, Kubera is known as Jambhala and is often sculpted with a mongoose in his left hand – the mongoose, when pressed, disgorges streaks of wealth or rounded coins from its mouth. Similar artwork has been found in the Greco-Buddhist art of Gandhara, in Tibet, in Nepal (the god of Mahakala) and China.

The Egyptian mongoose was well known to classical authors, with accounts in the literature particularly focused on the combat between mongooses and snakes. The Greek

[Left] A 10th-century sandstone statue of Kubera holding a mongoose-shaped purse in his left hand (northern India; San Antonio Museum of Art © Zereshk, CC BY-SA 3.0, Wikimedia Commons)

[Right] An 18th-century Tibetan bronze statue of Jambhala sitting on a snow lion and holding a mongoose in his left hand (© Clemensmarabu/Public Domain, Wikimedia Commons)

philosopher Aristotle wrote that 'the Egyptian ichneumon, when it sees the serpent called the asp, does not attack until it has called in other ichneumons to help; to meet the blows and bites of their enemy the assailants beplaster themselves with mud, by first soaking in the river and then rolling on the ground'. The ancient Roman authors Pliny the Elder and Claudius Aelianus, and the Greek historian Strabo, made similar remarks, although these were sometimes embellishments of other accounts and often inaccurate. For instance, Aelianus made the startling claim that all mongooses are hermaphrodite (each individual having both male and female sex organs) and that they fight before mating to determine which will be male and which female.

Fables about mongooses are reflected in a number of classical tales from India. The best-known source of Indian animal stories is from the books or chapters of the *Panchatantra*, which were related by the learned Brahman, Vishnu Sharma, and possibly dates from as early as 100BC. The better known of the two mongoose tales is from Book Five and is called *Hasty Action* or *The Brahman and the Mongoose*. In one version of this tale, a son is born to a Braham family, and on the same day, a female mongoose living in the house gives birth to a male pup, but immediately afterwards, she dies. The Brahman mother raises the young mongoose along with her own son, feeding them both with her own milk. One day, she leaves her baby sleeping in the house and goes out to fetch water. While the child is alone, a cobra comes out of a hole in the room. The mongoose fights with the snake to save his human brother, and cuts the cobra to pieces. In joy, the mongoose then runs outside to greet the son's mother, but when she sees the blood on the mongoose's mouth, she thinks that he must have killed her baby. She throws a pitcher full of water at the mongoose, which kills him. The mother then finds her baby still sleeping in his room, and a cobra, cut into pieces, lying by his side. She realises her mistake and repents, crying in sorrow that she has killed her adopted son, the mongoose.

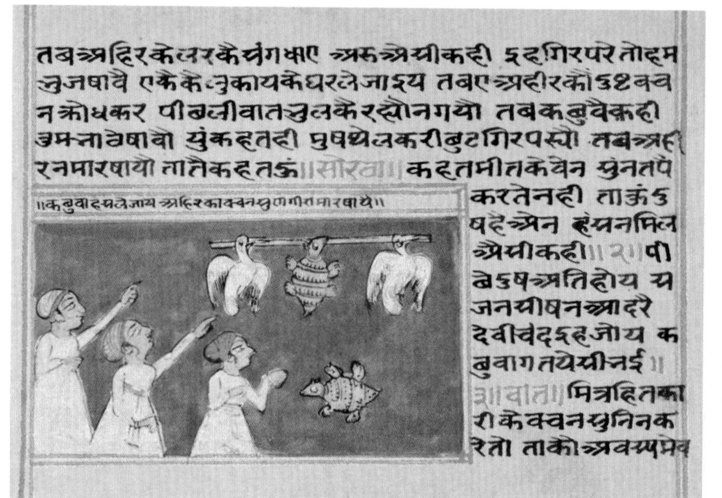

An 18th-century manuscript from Rajasthan, India, illustrating The Talkative Turtle, *one of the many animal fables in the* Panchatantra *(Stella Kramrisch Collection, Philadelphia Museum of Art, South Asian Art, Object Number: 1994-148-457/© Public Domain)*

Mongooses also appear in other stories in Middle and Far Eastern cultures and variations based on all these fables began to circulate in Europe from as early as the twelfth century. The mongoose is also mentioned in early Indian medical books, such as the *Charaka Samhita* (possibly around 100AD). An old Indian belief is that when bitten by a cobra, the mongoose goes to the jungle to look for a plant known as mungo root, which it eats as an antidote to the venom.

The traditional lore of the Malayan Aborigines, the Orang Asli, contains a story about the short-tailed mongoose and snakes. In this tale, the mongoose and a cobra once lived together in the jungle. The cobra is worried that his persistently hungry neighbour will one day eat up all the mice in the jungle, so he decides that he will get rid of the mongoose once and for all. One day, the cobra chances upon a sleeping python coiled up in a pile, and as the mongoose passes by, he notices the cobra examining this strange heap. The cobra grandly proclaims, 'This is the royal belt of the King Tikus Bulan and I have been ordered to guard it!'. The mongoose is very impressed, and begs to wear it around his waist. But the cobra teases the mongoose by constantly refusing his desperate pleads. After a while, the cobra concedes and allows the mongoose to put on the belt, but as he does so, the python awakes and immediately constricts itself around the unsuspecting mongoose. Thinking that the mongoose is dead, the cobra slithers away, hissing contentedly to himself. However, the mongoose does not die, and manages to escape. The mongoose and cobra are then mortal enemies from that day on. The mongoose learns how to evade the deadly strike of the cobra and to deliver a fatal bite, and later on, the mongoose vanquishes his foe in a fight to the death.

The English writer Rudyard Kipling granted the mongoose an everlasting place in literature when, in 1894, he wrote the short story of Rikki-Tikki-Tavi in the *Jungle Book*. In this tale, an English family, living in a big bungalow in India, discovers a young mongoose half-drowned from a summer flood and decide to keep it as a pet, which they name Rikki-Tikki-Tavi for his chattering vocalisations. One day, Rikki is confronted by two dangerous cobras, Nag and Nagaina. Nag plans to kill the human family and make Rikki leave, so that they can have a free run of the garden. Nag goes into the bungalow to kill the 'big man', but falls asleep in the bathroom while waiting for the family to wake up. Rikki grabs Nag by the head, and they thrash about furiously. This struggle wakes up the family and the father rushes into the room and fires a shotgun at the snake, blowing him into two pieces. Nag is thrown on the rubbish heap, where Nagaina mourns for him and vows vengeance. Rikki, well aware of her threat, searches for Nagaina's eggs and manages to destroy most of her brood. Meanwhile, Nagaina attacks the family at the dinner table and threatens to kill the family's son. Alerted to the crisis, Rikki races to his family with the last egg. Once there, Rikki shows the egg to distract Nagaina long enough for the father to pull his son to safety. Nagaina snatches her egg and flees to her hole where Rikki pursues her inside. After a long wait, Rikki emerges triumphantly, having killed Nagaina. The valiant Rikki then spends the rest of his days defending the family garden, where no snake would dare enter.

[Left] The first edition book cover of The Jungle Book (1894) by Rudyard Kipling, illustrated by John Lockwood Kipling (© Public Domain, Wikimedia Commons)

[Above] An illustration by John Lockwood Kipling from the first edition of The Jungle Book (1894) by Rudyard Kipling, showing the confrontation between the snake Nag and Rikki-Tikki-Tavi (© Public Domain, Wikimedia Commons)

As you can see, mongooses have been renowned for their ability to attack snakes for thousands of years, and even today, they can still be seen in fights with snakes, staged in Asian towns and villages as tourist attractions. A typical account of one is between a spectacled cobra and an Indian grey mongoose. At the start of the fight, the mongoose uttered a strident cry and walked up to the cobra with its tail bristling. They faced each other, and as the towering snake opened its jaws and drew back its hood to strike, the mongoose darted in and sprang for the lower jaw, simultaneously gripping the cobra's body with all four legs, as the mongoose bit the snake. The cobra writhed, sometimes taking the mongoose aloft with it, and as they struggled, the mongoose worked its jaws with a crunching action, its snout always keeping contact with the snake. This initial struggle lasted about five seconds, after which the mongoose broke loose. The cobra had been crippled by the mongoose's bites and could not raise itself to its former height; its lower jaw hung broken on the right side. The mongoose then made repeated attacks, each lasting about five seconds, during which it invariably targeted the snake's head, and not the body. In the early stages, the mongoose attacked with a quick rush from the side, but once it had slowed the cobra down, it sprang straight in regardless of danger. The mongoose would spring inside the cobra's striking circle and wait, sometimes up to eight seconds, until the dazed cobra opened its jaws to strike, when it would jump up, seize the jaw and continue biting. This particular fight lasted about thirty-five minutes and was then stopped, but had it continued it seemed certain that the cobra would have been killed. The mongoose had sustained two gashes in its upper lip, which were likely fang marks, but otherwise it showed no ill effects.

Written accounts of staged fights emphasise that it is the speed and agility of the mongoose that is chiefly responsible for its success. When the mongoose is fighting, its hair is erected so that it often appears twice its natural size, and this often causes the snake to strike short. But are mongooses immune to snake venom? Cobra venom generally contains neurotoxins that destroy nerve tissue, and the most common active ingredient, alpha-neurotoxin, works by attaching itself to acetylcholine receptor molecules on the surface of muscle cells – these receptors are designed to receive messages from nerves that tell the muscles to contract or relax. Alpha-neurotoxin blocks these messages, paralysing and ultimately killing the victim. Laboratory studies have revealed that mongooses have acetylcholine receptors that are shaped so that it is impossible for alpha-neurotoxin to attach to them. So, mongooses are resistant to cobra venom, but it remains to be discovered if they are also immune to viper venom, which typically contains hemotoxins that destroy red blood cells, disrupt blood clotting, or cause organ degeneration and tissue damage. Vipers seem to be much more effective against mongooses than cobras, and staged fights in the West Indies, between small Indian mongooses and vipers, more often than not end in the victory of the viper.

Recently, the meerkat has become extremely popular thanks to television documentaries and shows, cartoons, and blockbuster movies. The British television programme *Meerkat Manor* premiered in 2005 and ran for four series. This documentary-style programme was very well received by viewers and critics alike, and in 2007, it was

Animal Planet's top series, with an audience of more than four million in the United States alone. A companion book entitled *Meerkat Manor: The Story of Flower of the Kalahari* was published, and this series also led to the production of two feature films: *Meerkat Manor: The Story Begins* and *The Meerkats*. Meerkats have also featured in several Disney animated films and cartoon series, such as the character Timon from *The Lion King* franchise. The advertising campaign 'Compare the Meerkat', launched in 2009 on British and Australian television for an insurance comparison website, featured a fictional Russian meerkat called Aleksandr, which proved to be very popular with the general public: this became the fourth most visited insurance website in the UK, and resulted in a doubling in the site's overall sales. As of August 2009, Aleksandr had more than 700,000 Facebook fans and 22,000 followers on Twitter. The popularity of Aleksandr led to the publication of book in 2010, which reached no. 2 on the Amazon UK website in its first week of sale, and the release of other merchandise, such as cuddy toys, online downloads, and a smartphone application.

Banded mongooses have also been featured in a 2010 British television programme called *Banded Brothers*. Filmed and presented in the style of *Meerkat Manor*, it follows the regular lives and activities of a family of banded mongooses, which is being monitored in Uganda's Queen Elizabeth National Park by the Banded Mongoose Research Project of the University of Exeter in the UK.

Introduced mongooses

For centuries, humans have transported plants and animals from their native countries and introduced them to new locations, either accidentally, as when rats and domestic cats jumped from docked ships, or deliberately, often for controlling agricultural pests or simply for human enjoyment, such as the planting of exotic plants in parks and gardens. Unfortunately, many of these introduced species became invasive and had a devastating impact on their new environments, beyond what was originally intended. Whereas animals and plants in their native range are subject to control mechanisms, such as predators and diseases, which regulate their population numbers, these mechanisms are often lacking for introduced species in their non-native environment. Consequently, the numbers of some invasive species can sky-rocket, which may have harmful impacts for local wildlife. Even if the numbers of invasive predators do not reach high levels, native animals are often unable to defend themselves against these novel aliens, and may also not be immune to any diseases that they carry. Invasive species are now the second major cause of biodiversity loss across the world.

Rats and snakes are often pest species for humans: rats eat the crops of farmers and venomous snakes pose a danger to farm workers and the local people. On many small islands in the Caribbean, European sugar cane farmers were plagued by rats and snakes, and as there was a lack of native predators, exotic carnivore species were introduced as 'biological control agents'. Since early human times, mongooses were known to be efficient predators of small mammals, including rats, and were also renowned for their ability to kill snakes. This particular animal, then, seemed an ideal candidate, and one species was chosen: the small Indian mongoose from India – perhaps this species was singled out because Europeans had colonies in both India and the Caribbean. Once this mongoose had successfully been introduced to sugar cane plantations on some of the islands in the Caribbean archipelago, it was then transferred to other places. From the 1860s onwards, small Indian mongooses have been successfully introduced to over 60 islands around the world, as well as parts of mainland Europe and South America. There were failed attempts to introduce small Indian mongooses to Australia for controlling rabbits, and a few mongooses sighted on mainland Florida were captured through intensive trapping.

The introduced small Indian mongooses did their job well and helped to control pest species, and in doing so, they thrived in their new environments. Unfortunately, they also captured and ate many rare and endemic mammals, birds, reptiles, and amphibians. These native prey species had never before seen such fierce and voracious predators, as there were none previously on most islands where the mongooses were introduced, so they were unprepared for what hit them. Together with other introduced species, including rats and domestic cats, small Indian mongooses have caused the extinction of several animals, and many others are currently threatened.

Attempts have been made to eradicate the small Indian mongoose on several islands through trapping and poisoning campaigns; some people have even advocated hunting mongooses for sport, or that local inhabitants should eat mongooses as a meal! Despite all these killing methods and suggestions, this species has only been eliminated on six small islands. For example, the small Indian mongoose was introduced intentionally on Fajou Island (a small island off Guadeloupe) in the 1930s, but was successfully eradicated in 2001, after several trapping campaigns. The elimination of the mongoose, coupled with the control of introduced rats, have stopped the destruction of hawksbill turtle nests and the decline of the clapper rail (a chicken-sized bird that can rarely fly), and terrestrial crab populations are now recovering.

In 2010, a couple of small Indian mongooses were discovered within the port of Noumea, on the Pacific island of New Caledonia. Fortunately, they were immediately trapped – had a population of mongooses established itself here, it could have spelled a major disaster for the endemic fauna on this island. This incident highlights that the risk of new invasions still exists nowadays, whether intentional or not.

As well as being introduced for pest control, small Indian mongooses are sometimes imported to be kept as pets, which may escape or be let loose, either when they prove too difficult to handle or because the owners fear being prosecuted for importing an illegal animal.

A couple of other mongoose species have been introduced to non-native places. The Indian grey mongoose, from the Indian region, was introduced to Peninsular Malaysia (and possibly Sumatra), but it has not been recorded since the beginning of the 20th century and so seems to no longer exist there. This species was also released in central Italy during the 1950s, but for unknown reasons, their numbers decreased in the early 1980s, and they were considered extinct by 1984. A population of the Indian brown mongoose was discovered recently on the island of Viti Levu, in Fiji, and may have derived from a pair of individuals that were kept in a private zoo in the 1970s. The Indian brown mongoose now coexists on Viti Levu with the small Indian mongoose, which was also introduced to this island. The potential impact of these two species on the local fauna is extremely worrying, especially as one is nocturnal and the other is diurnal, which may increase the range of prey that is taken. Intensive efforts to eliminate them, or at least control their numbers, are being made.

The Egyptian mongoose, which mainly occurs in Africa, was thought to have been introduced into Portugal and Spain, but recent genetic studies have suggested that its presence in southern Europe may be the consequence of a natural dispersal during the lower sea-level of the late Pleistocene, around 700,000 years ago.

4. Researching mongooses

Only a few mongoose species have been intensively studied, which accounts for our general lack of knowledge about these animals. There are a wide variety of research techniques and methods that are available for scientists and naturalists to study animal species, and below we have described those that we, and others, have used for researching mongooses.

Direct observations

Some of the social species, such as meerkats and banded mongooses, live in open habitats and are active during the day, which allows field scientists to observe their daily behaviour and movements from a fairly close distance, especially if a group becomes habituated to the presence of humans. Obviously, this technique is not suitable for mongoose species that live in dense vegetation, such as rainforests, or are only active at night. However, nocturnal mongooses can sometimes be detected by slowly walking along a trail or logging road at night and scanning the undergrowth with a powerful torch or spotlight to pick up the bright, reflective eyeshine of an active animal. Such information can confirm their presence in a region, or in some specific habitat.

Radio-telemetry

Since the mid-20th century, field scientists have been using radio-telemetry to track the movements of animals in the wild. This technique involves two devices: one that emits a signal and another that detects it. The first device is a transmitter that is attached to an animal and which sends out a signal in the form of radio waves, just as a radio station does. For most carnivore species, this transmitter is attached to a collar that is placed around the animal's neck. The second device comprises a receiver and directional antenna that picks up the signal, just like a home radio picks up a station's signal, and this receiving system can be carried by a field scientist while walking around the study area, or placed in a vehicle or plane. Global positioning system (GPS) transmitters have now been developed that use a network of orbiting satellites to automatically record the locations of an animal at specific time intervals, which can then be downloaded to a computer or other mobile device. This new technology produces a large amount of data, and greatly minimises the time that researchers need to spend in the field.

Radio-telemetry can be used to yield valuable information on a mongoose's movement patterns, the size and use of its home range or territory, which types of habitat it prefers, its pattern of activity, and its choice of rest or den site. However, this is an invasive technique, in that to attach a radio collar to a mongoose, it must be trapped and anaesthetised.

[Above] In Krau Wildlife Reserve, in Peninsular Malaysia, this short-tailed mongoose was trapped for a radio-telemetry project. The animal was radio-collared and then released back into the forest (© Géraldine Veron and Andrew P. Jennings)

[Upper right] Field researcher Andy Jennings with an anaesthetized short-tailed mongoose. The animal was radio-collared, measured and weighed, and hair samples were taken for genetic analyses. The black mask around the animal's face protected its eyes while it was under the anaesthetic (Peninsular Malaysia © Géraldine Veron and Andrew P. Jennings)

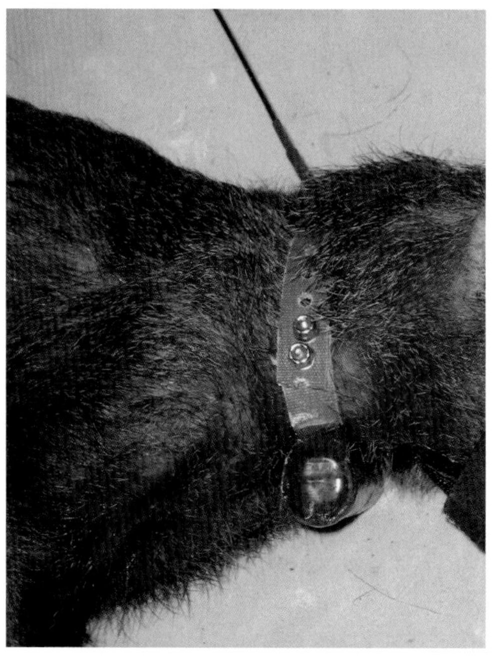

[Right] A radio-collar attached around the neck of a short-tailed mongoose (Peninsular Malaysia © Géraldine Veron and Andrew P. Jennings)

A white-tailed mongoose with a radio-collar in KwaZulu-Natal, South Africa, as part of a field project run by Jarryd Streicher (© Jarryd Streicher)

Camera trapping

Camera trapping is a method for capturing images of wild animals in the wild, and has been used in ecological research for several decades. A camera trap is a remotely activated camera system that is triggered by a motion and an infrared sensor – as an animal passes in front of the camera, images are automatically recorded. Originally, cameras with print film were used, but these have been replaced with digital models. Camera traps are often attached to trees, and placed close to trails, or near sites that may attract animals, such as a waterhole or animal carcass. Many camera traps can be set up across a study area and left in place for weeks or months, only requiring periodic checks to download images and change spent batteries.

Camera traps are particularly useful for detecting rare and elusive species, and a great advantage is that they can record a large amount of data with very little disturbance to the local animals. Camera trapping can be used to detect which mongoose species are present in an area, record their behaviour and activity, and indicate which habitats a particular species prefers. If individuals can be identified on images using their distinctive body patterns, it is also then possible to estimate a species' population size within the study area.

[Left] A camera trap attached to an oil palm tree in Sumatra. The aim of this study was to survey for small carnivore species within oil palm plantations in Indonesia (© Géraldine Veron and Andrew P. Jennings)

[Right] A camera trap photograph of an Indian brown mongoose in Brahmagiri Wildlife Sanctuary, India (© Devcharan Jathanna)

COMPUTER MODELLING

Up until now, not many mongooses have been studied in the wild, and for several species, we only have a few records of where they have been found. Over the past decade or so, computer modelling techniques have been developed that can use this limited field data to learn more about a species' distribution and ecological niche. One such technique is called ecological niche modelling and uses a computer program to combine location records with environmental information to predict where the ecological conditions are the most favourable for the species. These distribution maps can be very useful for conservation purposes, and for determining where future field surveys and research studies should be focused. Modelling analyses can also be used to make predictions for the future, such as the impact of climate changes.

MOLECULAR TECHNIQUES

Recent genetic studies have shed much light on the taxonomy and systematics of mongooses. Small samples of hairs, tissue and blood are taken from mongooses in the field (either from trapped animals, road kill, or hair traps), or small amounts of material are removed from the bones or skins of museum specimens. These samples are treated with chemicals to extract the DNA. Fragments of genes that are obtained from this DNA are amplified, and then sequenced, using a sequencing machine, to determine the base pair sequence of the genes. Computer programs can then compare the sequences of genes between individuals and species, and determine the genetic relationships between different individuals, populations or species. This helps trace their evolutionary history.

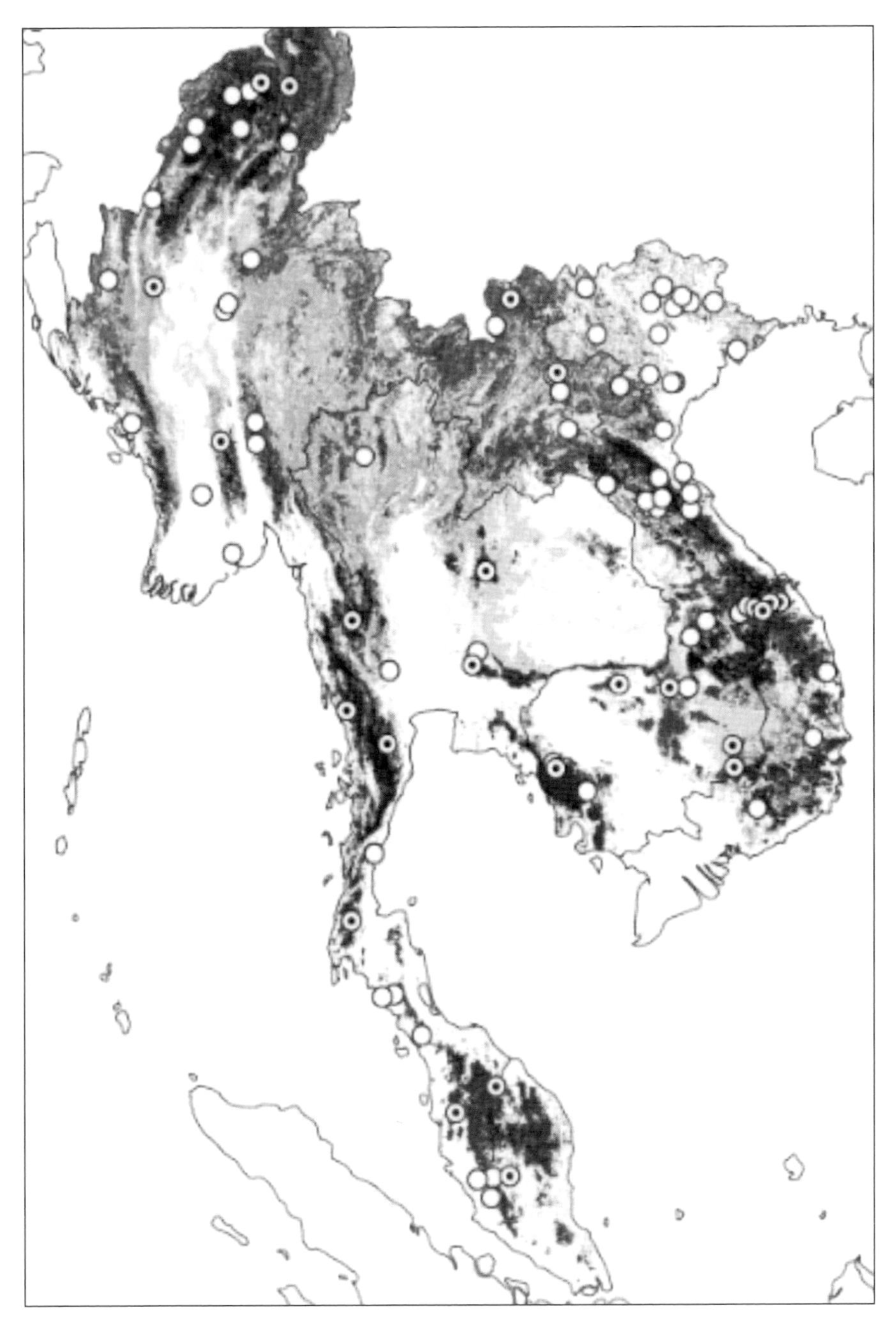

This map is an example of the output from an ecological niche modelling analysis on the crab-eating mongoose in southeast Asia. The dots mark the localities where this species was recorded and the grey shading indicates where the crab-eating mongoose is predicted to occur within this region, which roughly corresponds to the distribution of rainforest (© Andrew P. Jennings and Géraldine Veron, modified from: Jennings, A.P. and Veron, G. 2011. Predicted distributions and ecological niches of eight civet and mongoose species in southeast Asia. Journal of Mammalogy 92: 316–327)

5. Conservation

As predator species, mongooses play a crucial role in the various ecosystems that they occupy, upon which we depend for our own existence, so protecting mongooses benefits not only them, but us as well. But regardless of their usefulness to humans or ecosystems, all mongoose species have an inherent right to live and survive, and making people more aware of mongooses, and the threats that they face, will greatly help in preserving them for future generations. It is quite sobering to learn how little we know about most mongoose species: the Liberian mongoose was only discovered by Western scientists in 1958, Ansorge's cusimanse was known from only two specimens until 1984, and the mongoose species on Palawan Island, in the Philippines, was wrongly identified until 2015.

There seems to be a general perception among some scientists that mongooses have less demanding requirements for where they can live than their larger cousins, such as tigers and bears, and that mongooses are more tolerant of any destruction to their habitat and of any human presence. This may be the case with a few wide-ranging mongooses that can live in many different habitats, and close to towns and villages, but because of our general lack of knowledge about most species, we really do not know if this is true of all mongooses. Some mongoose species have restricted ranges, and many may have very specific habitat requirements, such as lowland forest or wetland areas, which would make them very vulnerable to human-induced environment changes, as well as other activities, especially hunting.

The global organisation *International Union for Conservation of Nature* (IUCN) regularly assesses the conservation status of animal and plant species worldwide, by evaluating their ecological requirements and the impact of any threats. These assessments are then published online on the IUCN's Red List of Threatened Species website (www.iucnredlist.org). The aim of these assessments is to ascertain the extinction risk of species, and to highlight to the general public and policy makers, such as government institutions and conservation agencies, which ones are in most urgent need of conservation action. Wildlife experts use precise criteria to evaluate the extinction risk of thousands of wildlife species (and some subspecies) by assigning a risk category to each one, which from highest to lowest are: Critically Endangered, Endangered, Vulnerable, Near Threatened, and Least Concern; if there is not enough information available for evaluating if a species is at risk, it is listed as Data Deficient.

The most recent IUCN assessments of mongooses, published in 2016, classified 29 species as Least Concern (i.e. at the lowest risk of extinction). Only the Liberian mongoose and the Sokoke bushy-tailed mongoose were listed as Vulnerable; Jackson's mongoose, short-tailed mongoose, and collared mongoose were categorised as Near Threatened, and Pousargues' mongoose was listed as Data Deficient. So it appears that almost all mongooses are not threatened with extinction, which would be great news, except that we believe that the conservation status of most mongoose species is currently too difficult to assess with any certainty, as too little is known about their population numbers and the precise impacts of any threats. At the moment, wildlife experts largely have to speculate about what is actually happening in the wild in order to assess the conservation status of mongooses, and despite the fact that most species are currently listed as Least Concern, the population trend of 22 mongoose species is either unknown or thought to be declining. In fact, one of biggest problems for mongoose conservation is our lack of knowledge of their distribution and ecology, which highlights that more field studies on mongoose species are urgently needed. We cannot develop and implement effective conservation actions unless we have more reliable information – so any students and field researchers reading this and looking for ideas for new projects, please take note.

Habitat loss

One of the major threats that mongooses face is the destruction of their habitat. Many natural habitats are fast disappearing due to agricultural expansion, mining activities, road building and the expansion of cities, towns, and villages. Forest-dependent mongoose species are particularly threatened – up to 90% of West Africa's coastal rainforests have disappeared since 1900, and across South Asia about 88% of the rainforests have been lost. World annual deforestation is estimated at around 14 million hectares a year, equal to the area of England, with the highest rates occurring within tropical rainforests.

Mongoose species that need forests in which to live will disappear if all the trees in an area are cut down. Even if some forest patches are left intact, the breaking up of a large expanse of forest into smaller fragments can still have harmful impacts for forest mongooses. Forest fragmentation divides a species population into smaller units, which then have a higher risk of going extinct – an outbreak of disease, or a forest fire, can more easily wipe out a small, isolated population. If a mongoose species is unable to cross open areas between forest patches, this will restrict the dispersal of individuals between isolated populations, resulting in a higher level of inbreeding that can lead to harmful consequences. Even selective logging, although not as destructive as clear-cutting, can dramatically alter the overall structure of a forest, and this may affect the availability of food resources and suitable den sites. The network of roads that are created within a logged forest allows greater access to hunters, which will lead to increases in hunting pressure.

To counter the loss of forest habitat, remaining forests must be protected and conserved, and forested corridors connecting existing forest patches need to be maintained. However, it is unlikely that all forests will be granted a fully-protected status, and logging activities will continue in commercial forests, so the management of non-protected areas also needs critical analysis in order to incorporate the requirements of mongooses and other wildlife species.

Some mongooses are found in wetland areas, and along the banks of rivers and streams. These species are highly vulnerable to the drainage of swamps and marshes, and the pollution of watercourses, and could be severely impacted unless these water bodies are protected, or at least maintained in ways that minimise any harmful effects.

[Left] An area of cleared rainforest in Peninsular Malaysia, with recently planted oil palm trees. Across southeast Asia, rainforests are being cleared at an alarming rate to create farmland and oil palm plantations, and this is having severe impacts on forest-dependent mongoose species (© Géraldine Veron and Andrew P. Jennings)

[Right] A young oil palm plantation in Peninsular Malaysia in an area that was formerly rainforest. Cultivated land has replaced large portions of forest in many lowland regions, with small remnant forest patches only left on hills and in swampy areas which are too difficult to clear and cultivate (© Géraldine Veron and Andrew P. Jennings)

[Below] The expansion of oil palm tree plantations in Malaysia and Indonesia threatens rainforest biodiversity in southeast Asia. The creation of new roads also makes it easier for hunters to access remote areas (Peninsular Malaysia © Géraldine Veron and Andrew P. Jennings)

Hunting and wildlife trade

Numerous animal species are hunted for food and to supply the huge illegal trade in traditional medicines, skins, bones and pets, which has been estimated at between US$10 and $20 billion per year. Although this wildlife trade is global, it is most acute in China and across Southeast Asia. Wildlife traders penetrate into the remotest areas and actively encourage the hunting of species for which there is a demand. Animals are killed using various means, including guns and wire-snares, and since many types of traps are unselective in what they catch, non-targeted animals are also taken and killed. Cutting forests increases the risk of hunting by creating access roads to remote areas.

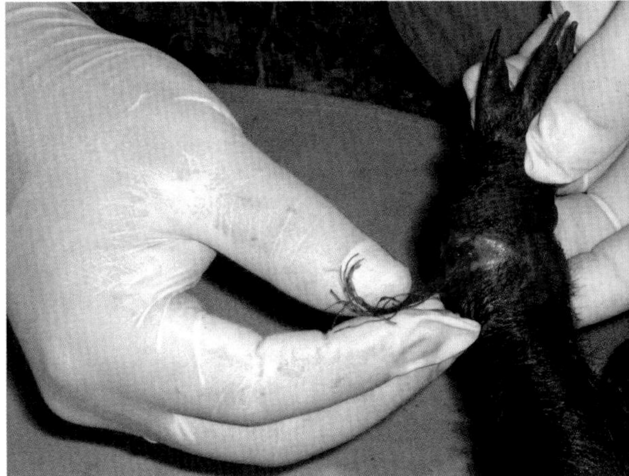

[Left] Various live animals (including the endangered pangolin) for sale in a wildlife market in Möng La, Shan, Myanmar (© Dan Bennett, CC BY 2.0, Wikimedia Commons)

[Right] The leg of this short-tailed mongoose had been caught in a snare trap. The cord around the leg had tightened and embedded itself within the muscle, causing both the leg and foot to swell. Fortunately this mongoose had managed to escape before the hunter had reached the snare trap – other animals are less lucky and suffer greatly as they die. We were able to remove the embedded cord and treat the wound. This animal was then radio-collared and tracked for several months, which indicated that it had recovered well (Peninsular Malaysia © Géraldine Veron and Andrew P. Jennings)

Mongooses and other wildlife are a source of meat (usually called bushmeat when people eat wild animals) for local communities across Africa and Asia. The amount of wild meat eaten by some African communities is quite substantial, particularly where the breeding of domestic animals is problematic because of trypanosomiasis and other parasites; for instance, in Ghana, bushmeat can make up 62% of the animal protein that is eaten, with small carnivores (including mongooses) making up to 15% of all the bushmeat

Bushmeat for sale at a roadside stall in Ghana (© Wikiseal CC-BY-SA-3.0, Wikimedia Commons)

consumed in some communities. In the Democratic Republic of the Congo, Ansorge's cusimanse is the most frequently killed mammal, accounting for 6% of all mammal bushmeat, and five mongoose species have been found in bushmeat markets within southeastern Nigeria.

Across Asia, humans have been hunting wildlife in tropical forests for food and traditional medicines for over 40,000 years, but over the last hundred or so years, there have been significant advancements in hunting methods, such as the use of guns, and a huge explosion in demand for wildlife products. While bushmeat was once consumed primarily by poor and rural communities, it is increasingly finding its way into more affluent urban centres, particularly in countries such as China. Wildlife harvesting has now reached unsustainable levels due to a burgeoning human population and shrinking forests, and commercial poaching to supply the ever growing illegal wildlife trade is now a major threat to the biodiversity of East, South, and Southeast Asia. Mongooses are seen in both local and regional bushmeat markets throughout Asia, but we do not know the full scale of the problem, nor to what extent the current levels of hunting pressure are impacting local populations.

There is not a large demand for mongoose fur for the fashion industry, unlike the fur trade in mink, sable and wild cats. Although skins of mongooses are seen in village and town markets, it appears that they are mainly sold for local uses, such as making shaving brushes or as good luck charms. However, mongooses in India are killed in very large numbers so that their coat hairs can be turned into paintbrushes, which are then sold by domestic and international art and craft suppliers, particularly in America and Europe. Indigenous hunters typically trap mongooses using snares or nets, and then beat them to death with clubs. Hunters pull the fur off the skin, keeping the meat for themselves, and then sell the hair to middlemen. Producing mongoose paintbrushes is a large-scale operation: in the early 2000s, some 50,000 mongooses were being killed annually. In 2015, Indian law enforcement seized 14,000 mongoose-hair paintbrushes from a distributor in a coastal town of southwestern India, and in August 2016, Indian officials arrested people suspected of smuggling 12 pounds of mongoose hair, the equivalent of more than 130 animals. Despite a 1972 Indian law banning the killing of mongooses, and several recent crackdowns by law enforcement agents, this illegal trade continues in India. There is a ban on the international trade of Indian mongooses, and they are listed under CITES – the Convention on International Trade in Endangered Species of Wild Fauna and Flora, a global treaty that regulates the cross-border wildlife trade – yet mongooses are still being killed and traded. Unfortunately, there is little knowledge among artists and buyers about this cruel practice, and generating awareness among consumers is one of the biggest challenges to stopping this barbaric trade. But efforts are being made. In 2002, the protection of mongooses was upgraded in India, and alternative paintbrushes have sprung up to replace those made with mongoose hairs. As brushes made of mongoose hair are often sold as sable or badger brushes to avoid legal hassles, being able to identify mongoose hair in brushes will play a key part in curtailing this illegal trade. To aid in this endeavour, the Wildlife Institute of India has recently published manuals for identifying mongoose hairs, using banding patterns and microscopic analyses.

Mongooses are also taken from the wild and kept as pets, either as companions or to protect houses from dangerous animals, such as snakes and scorpions, and to reduce mice and rat populations. Mongoose skulls found in houses at Merkes, Babylon (now in Iraq) suggests that even in these early times (around 600 BC) mongooses were kept as household pets. The mongoose species that are most readily tamed are those that are gregarious – their need for social contact is so great that they quickly come to accept a human as a substitute companion. However, some of the solitary species, such as the small Indian mongoose, are also said to make a pleasing pet if they are caught young enough; they often become attached to the people they know, and will attack strangers. Unlike dogs, though, and more like cats, they are quite independent.

Throughout Asia, there is still a large demand for and trade in animals as pets, and this includes mongoose species. Plantation and forest workers may supplement their income by opportunistically catching mammal species to sell as pets (and for other uses) in local markets or to wildlife dealers, and hunters may specifically target animals that are in high demand. Most of this wildlife trade is illegal, and local and international wildlife laws are usually ignored and not enforced. Medan, in northern Sumatra, is one of several major centres in Southeast Asia for both domestic and international wildlife trade. A five-year study of the pet industry here recorded 300 bird species, 34 species of mammals, and 15 reptile species. The Javan mongoose was frequently observed for sale, with a total of 324 animals recorded during the monthly surveys; obviously, the actual number sold over the

In many markets and restaurants throughout Asia, many animals are displayed alive in cages to be sold for food or as pets (common palm civet, Vietnam © Géraldine Veron)

five-year period was far greater than this figure. Many of these animals are exported by sea and air to Singapore, Malaysia, Thailand, China and other destinations around the globe.

Obviously, we do not recommend that anyone should keep a mongoose as a pet; to do so would fuel the rampant illegal wildlife trade and encourage local hunters to trap animals, leading to the needless suffering and deaths of many individuals (for every animal trapped and transported, many more die in the process), as well as causing further declines in mongoose populations in the wild. It is infinitely better to admire and marvel at mongooses in their natural habitat than to keep one in your home, and there are many domestic animals that would make more suitable pets.

Persecution

Some mongoose species prey opportunistically on domestic poultry, and many sheep farmers believe that mongooses are important predators of their lambs, although this is based more on myth than fact. Whether the threat is real or not, though, mongooses are often killed by local people if it is believed that this will reduce poultry or livestock losses, and this may be why the number of Egyptian mongooses around North African villages has decreased in recent years.

Mongooses are also persecuted because of their possible role in the spread of rabies and other diseases, such as leptospirosis, which was recently discovered in banded mongooses. Rabies is known to occur in several mongoose species, including the yellow mongoose and small Indian mongoose. However, there are only a few documented cases of rabies-infected mongooses attacking humans or domestic animals. Although some mongooses may pose a health risk in areas where they come into close contact with humans, vaccinating mongooses in the wild could be a better alternative option to killing them.

Climate change

Climate changes can alter the distribution of weather patterns and global temperatures, and these may last from decades to millions of years. There are many examples of natural causes, including variations in the solar output, volcanism, and plate tectonics (the movement of continental landmasses). These climate changes can have many profound impacts – the recent Ice Ages are a well-known example – that may result in changes in the sea level and the distribution and coverage of vegetation, such as forests. Animals respond to these changes in one of three ways: they either adapt to the new local conditions, disperse to where their native habitat has shifted, or go extinct.

There is now compelling evidence that humans are causing climate changes, and this is happening at such an unprecedented rate and scale that most animals, including mongooses, may not be able to adapt rapidly enough, or may be prevented from dispersing to where remaining areas of their preferred habitat still exist. The consequence of human-caused climate changes is that extinctions are happening at far higher levels than ever previously experienced in the history of life on earth, which is not only causing a massive loss in biodiversity (potentially including many mongoose species), but threatens human existence as well.

Conservation actions

The limited information we have about the natural history of most mongooses makes it very difficult to formulate good conservation strategies, so one of the most important actions that can be done to help preserve them is to undertake scientific studies. We hope that field biologists will focus more research effort on mongooses across the world.

Since the loss of their habitat is one of the major threats to mongooses, habitat preservation is crucial for their long-term survival. Many national parks and reserves have already been set aside for wildlife; however, the total area of protected land is currently only about 10–15% of the world's land surface area. Most protected areas are probably too small in themselves to support mongoose populations, and are often surrounded by a sea of human-created habitats, such as farmland and plantations, which are hostile to most mongooses and may also prevent them from dispersing from one favourable area to another. So although we should strive as much as possible to preserve new areas for mongooses, and other wildlife, it is unlikely that we will be able to protect sufficient land for all species. This means that the way non-protected land is managed will also have a crucial impact on mongooses. Making agricultural land more wildlife-friendly, or at the very least more amenable as a travel corridor from one favourable habitat to another, will play a crucial role in developing conservation strategies for mongooses across rural landscapes. Climate changes will also compound this situation. Global changes in temperatures and rainfall patterns will likely cause the existing distribution of habitats to shift, and some mongooses may go extinct if they cannot adapt to these rapid changes, or are not able to move to new areas. Many non-government organisations (NGOs), both at the local and global level, are working hard to preserve animals and their habitats, and we would encourage greater focus and attention on the lesser known species, including mongooses.

We know that some mongoose species are hunted for bushmeat, for their hairs, or to be kept as pets. Education and public persuasion are key tools for stopping people from taking mongooses from the wild, and to combat these threats, we need to raise awareness about these issues, encourage alternative practices, and implement and enforce effective laws and regulations.

The captive breeding of animals is often promoted as a conservation tool, especially for species on the brink of extinction. Animals have been kept in zoos, menageries or private collections since ancient Egyptian times. Exotic species were taken from the wild and displayed as curiosity objects to the general public, or they were often used by monarchs and rulers as symbols of power. In more recent times, zoos have increasingly became more focused on presenting exhibits that educate, as well as entertain, the general public. A primary goal of many zoos today is to breed endangered species in captivity, as well as to provide research and educational opportunities and a source of inspiration for conservation actions. Some zoos also support in-situ research and conservation programmes by providing funds and expertise for field projects on threatened or poorly known species. However, many critics and animal rights activists consider that zoos are immoral and serve only to fulfil human leisure desires, at the expense of the animals. Keeping animals in captivity also raises many ethical and welfare issues, especially in poorly managed zoos, and many conservationists argue that captive breeding should be seen as supporting, rather than replacing, conservation efforts in the wild, and perhaps should only be used as a last effort to save a species. Captive breeding over several generations may unintentionally select animals that would not be capable of surviving in the wild, and this could result in the failure of any reintroduction programmes. This is particularly problematic for carnivore species, which have complex social behaviours and must learn how to find and capture their prey.

At one time or another, many of the 34 mongoose species have been kept in zoos throughout the world, but the most frequently exhibited species tend to be the social mongooses, such as the meerkat, common dwarf mongoose, banded mongoose, and yellow mongoose. This is not too surprising, as these particular mongoose species are gregarious, very active during the day and fun to watch, which makes them particularly appealing to many zoo visitors. The captive breeding of many endangered species is now coordinated by cooperative breeding programmes, such as a Species Survival Plan (SPP) or European Endangered Species Programme (EEP). Mongooses are treated in a *Regional Collection Plan* and *Mongoose, Meerkat, & Fossa (Herpestidae/Eupleridae) Care Manual*, both published by the Small Carnivore Taxon Advisory Group of the Association of Zoos and Aquariums. Currently, there are programmes for the meerkat, banded mongoose and common dwarf mongoose, although these three species are not considered threatened, according to the most recent IUCN Red List assessments. The one mongoose species that is currently listed by the IUCN as endangered, the Liberian mongoose, is not in any zoo breeding programmes. The Liberian mongoose was only discovered by Western scientists in 1958, and has a very restricted range in West Africa, so it is not too surprising that this species has not been kept in captivity (except for a single male Liberian mongoose that was captured in 1989 and held in Toronto Zoo) nor that a zoo breeding programme is yet to be set up. As we gather more field information about the conservation status of all mongooses, the incentive for zoos to establish breeding programmes for any that are considered threatened may change in the near future.

A few mongoose species are displayed in zoos around the world (yellow mongoose; Ménagerie du Jardin des Plantes, Paris, France © Géraldine Veron)

HOW YOU CAN HELP TO SAVE MONGOOSES

There are many ways you can help to preserve mongooses for future generations and here we have suggested a few examples:

- You could go into the field to study mongooses. There are tremendous research opportunities for a field scientist to study the ecology of mongooses in the wild. If you are not a professional biologist, you could volunteer on a field project that is studying mongooses (see below for a couple of examples).
- Devote some of your free time to help raise awareness about mongooses and the conservation issues they face through using social media and dedicated websites.
- Campaign for more effective legislation and actions that will protect mongooses from hunting and wildlife trafficking, and preserve their habitats.
- You can make good shopping choices that will have powerful knock-on effects for the benefit of mongooses (and other wildlife). We are all consumers and the decisions on what we buy each time we go shopping can ultimately have profound impacts on the environment. Buy only wooden products that are made from trees grown in sustainably managed forests, for instance, and avoid those containing exotic woods taken from tropical rainforests. Boycott products containing palm oil to discourage the expansion of oil palm plantations, one of the major causes of deforestation in tropical regions.
- Support NGOs and field research projects that are either directly focused on mongooses, or are striving to protect the habitat in which they live. Many of these

have online websites that can easily be found by searching the internet. Here are two examples:

- The Banded Mongoose Research Project <http://socialisresearch.org/about-the-banded-mongoose-project/> This is a field project in East Africa that is studying the social behaviour of banded mongooses and is run by a team of researchers from Uganda and the University of Exeter.
- The Dwarf Mongoose Research Project <https://dwarfmongooseresearch.weebly.com/> This is a field project studying the social behaviour of common dwarf mongooses and is run by the University of Bristol, UK and the University of Pretoria, South Africa.

Species accounts

This section contains species accounts for all the 34 species of mongooses, and is arranged in alphabetical order, first by genus, and then specific name. For each species, we have provided the most up-to-date information about its natural history; unfortunately, there is little information for many mongoose species, as only a few have been intensively studied in the wild. We have also included a colour picture, which for most mongooses was taken by an automated camera trap set up in the field to survey for animal species. No pictures of the Somali slender mongoose and Ansorge's cusimanse are currently available, so we have used illustrations to depict these two species.

The conservation status and distribution map for each mongoose are based on the IUCN species accounts that were published online in 2016 (www.iucnredlist.org). Our knowledge of a mongoose's distribution is generally quite poor, so the ranges shown for many species (in green) are approximations of the real situation in the wild. You should also bear in mind that many mongoose species occur in specific habitats, and so their actual distribution could be very patchy, depending on where their preferred habitat is found within its range, and how much remains in its original state.

GENUS *ATILAX*

There is only one species in this genus, the marsh mongoose, which is found in Africa. It has also been placed in the genus *Herpestes*, but recent genetic studies have shown that this species should be placed in its own genus.

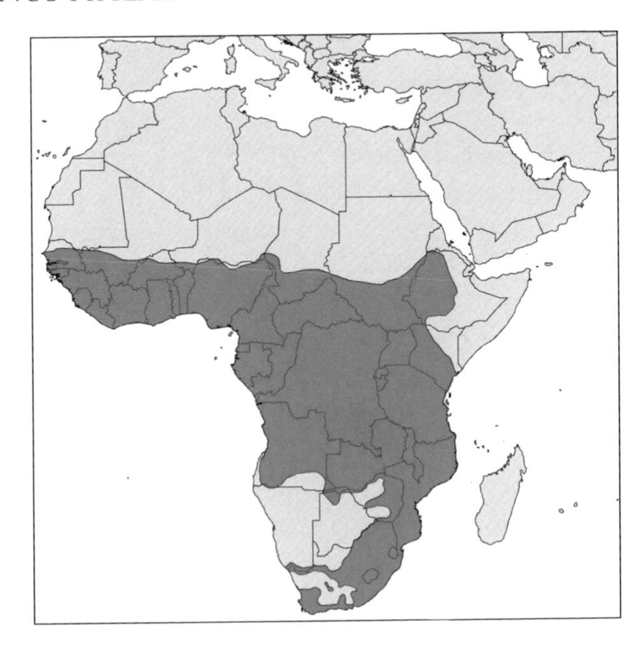

MARSH MONGOOSE

Atilax paludinosus

Also known as water mongoose

Total length: 79–97 cm
Weight: 2–4 kg
Red List status: Least Concern

Marsh mongoose (South Africa © Jarryd Streicher)

Marsh mongooses have a dark reddish-brown or black shaggy coat that has a grizzled appearance due to the lightly-coloured bands on each hair. Their fur is particularly long and thick in front of the ears, which help protects the inner ear from water. They have a long body and short legs. Their head is broad and triangular, the ears are small and rounded, and the feet are unwebbed, with five digits, naked pads and short, curved, non-retractable claws.

Marsh mongooses are found throughout most of Africa south of the Sahara, from sea level up to at least 2,500 metres. They live in both riparian and marine habitats – along streams and rivers, within swamps and marshes, and in estuarine and coastal areas. Marsh mongooses usually require dense vegetation and a source of water for survival, but they may sometimes wander away from water to search for food.

Marsh mongooses are active at night and during the twilight hours, and rest during the day within dense vegetation, or in burrows, on high, dry ground, close to water, using different daybeds each day. They are usually solitary, but during the breeding season may be seen in small family groups, consisting of a mother and her offspring. Home ranges are generally linear in shape, along watercourses, and can be up to 1.3 km^2 for males, and around 2 km^2 for females.

Aquatic prey forms the bulk of the diet, which includes crustaceans (especially crabs), insects, frogs, toads, molluscs, fish, small mammals, reptiles, birds, eggs, and sometimes fruits and carrion. Marsh mongooses mainly forage along the edges of water bodies, although they can also dive on occasion, and spend most of their time walking along the shore, or wading in water. They are quite dexterous and use their sensitive forepaws to forage in shallow water, reaching into mud, under rocks, and in crevices to search for prey. Any food items captured in water are usually carried out before being eaten. To break open hard-shelled prey, such as crabs and bird eggs, a marsh mongoose stands up on its hind legs and uses its forelimbs to smash the shell against a rock. Whereas small crabs are eaten whole, larger crabs are turned upside down and the fore claws bitten off; the meat is then eaten, and the carapace discarded. Rodents are seized in the jaws and flicked sideways before being given a killing bite, and insects are grasped in the mouth, or with the forefeet.

Marsh mongooses communicate with conspecifics through a variety of vocalisations, scent marking (using both cheek and anal glands), and behavioural displays. They defecate in latrines on rocks or sandy beaches, near streams and in marshy clearings, or under areas covered by vegetation. Marsh mongooses may eject a strong-smelling brown fluid from their anal glands when stressed.

Breeding seems to be seasonal in some regions and not in others. Gestation lasts 69-80 days, and the litter usually contains two or three pups, weighing around 100g at birth. The young are born blind and softly furred – their eyes open after 9–14 days. Only the mother raises the young, and during this time, the pups play, fight, and groom each other. The pups are weaned at 30–60 days, but they remain with their mother until they are able to fend for themselves, at about 20 weeks of age. In captivity, marsh mongooses have lived to just over 19 years.

The IUCN does not consider the marsh mongoose to be endangered, but populations in the wild are thought to be declining. Possible threats include habitat loss, water pollution, the drainage of swamplands to form arable land, and the hunting of marsh mongooses for bushmeat markets.

Genus *Bdeogale*

Mongooses in the genus *Bdeogale* occur in central and eastern Africa. The number of *Bdeogale* species is currently debated by taxonomists, with as many as four recognised by some authors. The bushy-tailed mongoose and black-legged mongoose are generally accepted as two distinct species. A subspecies of the bushy-tailed mongoose, which is found in the coastal forests of Kenya and northern Tanzania, is also considered by some scientists to be a separate species, the Sokoke bushy-tailed mongoose (*Bdeogale omnivora*), as it has a different coat colour and is slightly smaller than the bushy-tailed mongoose. Jackson's mongoose, which is found in East Africa, is treated as a subspecies of the black-legged mongoose by some authors, whereas others advocate that there are sufficient skin and skull differences for this mongoose population to be recognised as a separate species.

The species or subspecies status of both the Sokoke bushy-tailed mongoose and Jackson's mongoose have yet to be verified by genetic studies, using DNA samples taken from live animals or museum specimens, and hopefully future investigations will shed more light on the taxonomic status of these two taxa – of course, obtaining genetic data is far easier said than done for such rare and elusive animals.

Based on the current evidence, we have decided to recognise three *Bdeogale* species: the bushy-tailed mongoose, Jackson's mongoose, and the black-legged mongoose.

BUSHY-TAILED MONGOOSE

Bdeogale crassicauda

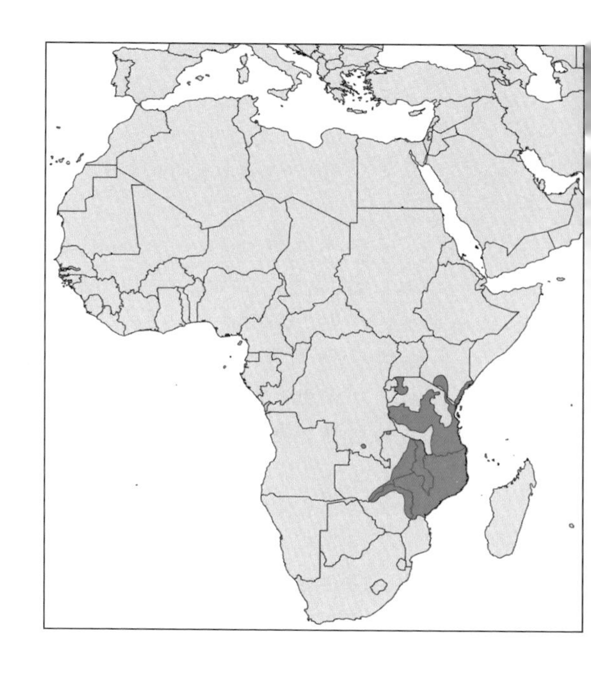

Total length: 58–80 cm
Weight: 1.5–2 kg
Red List status: Least Concern

Bushy-tailed mongooses appear black from a distance, but actually have a grizzled dark brown body coat, with black limbs and a prominent black bushy tail. The head is quite rounded for a mongoose, which gives the appearance of a short muzzle. The ears are short and rounded, and the eyes have a greyish-brown iris. The feet have four digits, with stout, curved claws.

Bushy-tailed mongooses are found in eastern Africa, including the island of Zanzibar, and live in a variety of habitats, including dense forests, woodland savannah, wooded grasslands and hilly areas with rocky outcrops, from sea level up to at least 1,850 metres. They are mainly active at night, and are usually solitary, although occasionally two or more animals are seen together, most likely a mother with her young. Bushy-tailed mongooses feed mainly on insects, particularly ants and termites, but will also eat other invertebrates

Bushy-tailed mongoose (Tanzania © Francesco Rovero)

(including millipedes, spiders, scorpions, and snails), small mammals, amphibians, and reptiles. On the island of Zanzibar, bushy-tailed mongooses have been seen eating large land snails, which they smash against coral outcroppings, stones, or tree trunks. Bushy-tailed mongooses use shaking and a neck bite to kill mice and snakes, and hold the dead animal between their forefeet to tear pieces off to eat. Spotted hyaenas and crowned hawk eagles are two predators that are known to catch and kill bushy-tailed mongooses.

Bushy-tailed mongooses are seldom seen and there are relatively few specimens in museum collections, so this species is thought to be rare throughout its range, and perhaps occurs in small isolated populations. However, some recent camera trapping studies in Tanzania have indicated that bushy-tailed mongooses can be relatively common in some areas. The IUCN does not consider this species to be threatened, as it has a fairly wide distribution, and lives in a variety of habitats. There are thought to be no major threats, but miombo woodlands across south-central Africa are being cleared for agriculture, or heavily disturbed through timber extraction. And, even though it is not known if bushy-tailed mongooses are specifically hunted for bushmeat, they are probably trapped as a by-catch in snares.

The IUCN recognises the Sokoke bushy-tailed mongoose as a distinctive species, and has listed it as Vulnerable, as it may have undergone a significant population decline over the past ten years due to the extensive and ongoing habitat loss in the coastal forests of eastern Africa.

JACKSON'S MONGOOSE

Bdeogale jacksoni

Total length: 79–90 cm
Weight: 2–3 kg
Red List status: Near Threatened

Jackson's mongooses have a grizzled, silvery-grey body coat, with dark brown or black legs, and a bushy white tail. The muzzle is brownish white and quite blunt, with a large nose, and the ears are round and broad. The cheeks, throat, and side of the neck are yellowish. The legs are quite long for a mongoose, and the feet have four digits, with thick, strong claws.

Jackson's mongooses live in lowland forests, bamboo forests and montane forests in parts of Kenya, Uganda and Tanzania, up to at least 3,300 metres. They are mainly active at night and usually solitary, although they are sometimes seen in family groups of two to

Jackson's mongoose (Tanzania © Francesco Rovero)

four. Jackson's mongooses mainly feed on rodents and insects, including ants and beetles, but will also eat other invertebrates, such as millipedes and snails, as well as lizards, birds, eggs, and carrion.

Jackson's mongoose appears to be rare and only occurs in isolated populations within its restricted distribution. Given its apparent dependence on forest habitat, the IUCN considers that this species is close to being endangered because of ongoing forest loss across its range. Hunting for bushmeat could also pose a serious threat for local populations.

BLACK-LEGGED MONGOOSE

Bdeogale nigripes

Also known as black-footed mongoose

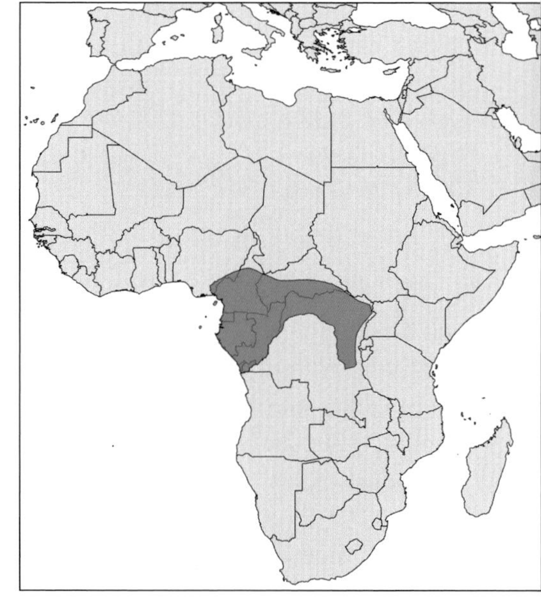

Total length: 75–104 cm
Weight: 2–5 kg
Red List status: Least Concern

Black-legged mongooses have a grizzled, silvery-grey body coat, light brown to brownish-black legs and feet, and a moderately bushy, white to creamy-coloured tail. The head is lighter coloured than the body, with a quite blunt muzzle, large nose and short rounded ears, which give the head a dog-like appearance. The legs are relatively long, and the feet have four slightly webbed digits.

Black-legged mongooses mainly live in undisturbed rainforest across Central Africa, from sea level up to at least 1,000 metres. They are mainly active at night, sleeping during the day in holes between the roots of big trees and in the dens of porcupines, and are usually

Black-legged mongoose (Gabon © Laila Bahaa-el-din)

solitary, except during the breeding season. Black-legged mongooses hunt mostly on the ground, and mainly feed on arthropods (chiefly ants, termites, beetles and grasshoppers) and small mammals (especially shrews and rodents), as well as snakes, lizards, frogs, toads, and occasionally some fruit; they also scavenge. Breeding may occur at the beginning of the dry season, and at least one or two young are born between November and January.

The IUCN does not consider the black-legged mongoose to be endangered, since it appears to be widespread in a region of relatively intact habitat, but numbers in the wild may be declining due to the ongoing destruction of forests and bushmeat hunting.

GENUS *CROSSARCHUS*

Mongooses in this genus are called cusimanses (or sometimes kusimanses), and there are four species that are found across West and Central Africa. They are generally small, shaggy-haired, short-legged mongooses, with distinctive long muzzles.

ALEXANDER'S CUSIMANSE

Crossarchus alexandri

Total length: 58–76 cm
Weight: 1–2 kg
Red List status: Least Concern

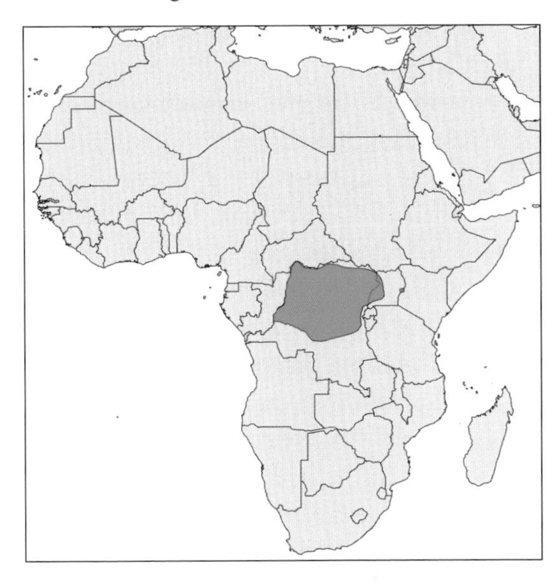

Alexander's cusimanses are the largest of the four cusimanse species and have grizzled, dark brown fur that is particularly thick and shaggy on the body, and is shorter around the face. They have a long, snout-like muzzle,

Alexander's cusimanse (Chinko Nature Reserve, Eastern Central African Republic © Thierry Aebischer and Chinko Research Team)

fairly large rounded ears, short legs, and a short tapering tail. The feet have five digits, with quite long, sharp claws.

Alexander's cusimanses are found in Central Africa (Democratic Republic of the Congo and Uganda) and live in lowland and montane rainforests, particularly amongst thick undergrowth and in swampy areas. This is a social mongoose, with as many as 10–20 animals in a group. Alexander's cusimanses are mainly active during the day, and will sleep together in holes within the trunks of dead trees. They spend most of their time on the ground, but can also climb sloping trees, especially when disturbed.

Alexander's cusimanses search for food along the forest floor, and use their long muzzle and strong claws to open rotting tree trunks. While foraging, they push their noses into crevices, scrape and scratch continually, and emit grunting and twittering contact calls. Alexander's cusimanses eat invertebrates, including earthworms, millipedes, slugs and insects (chiefly termites, crickets, grasshoppers, and caterpillars), as well as frogs, snakes, and some fruit.

When alarmed, a loud chatter is generated within the group, which will then run off in silence. To scent mark their territory, individuals will do a reverse handstand against a tree and smear the trunk with the smelly secretion from their anal glands. Captive animals tend to defecate and urinate in one place, and will superimpose their own urine and anal deposits over those of other animal species. Virtually nothing is known about their breeding behaviour, but according to local people, Alexander's cusimanses have three to four pups per litter.

The IUCN does not consider the Alexander's cusimanse to be endangered, as it is believed to be widespread and common in a region with relatively little-encroached habitat. However, populations in the wild are probably declining due to the destruction of rainforests and hunting for bushmeat.

Ansorge's cusimanse

Crossarchus ansorgei

Also known as Angolan cusimanse

Total length: 53–58 cm
Weight: 0.6–1.5 kg
Red List status: Least Concern

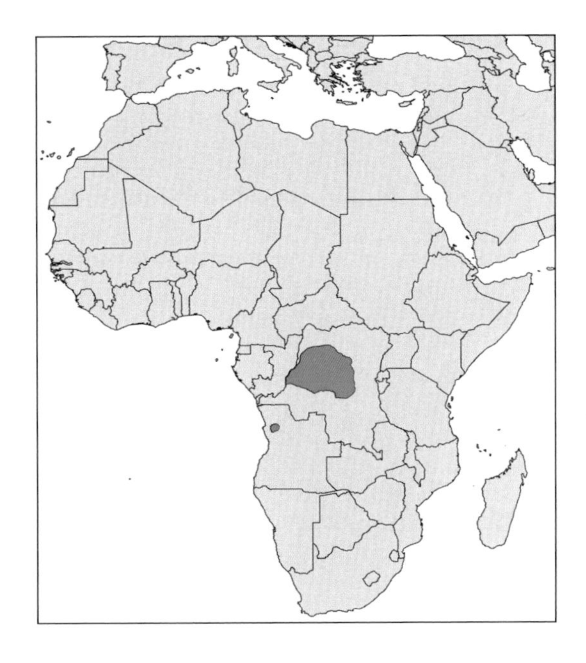

Ansorge's cusimanses have reddish-brown to black fur that is thick and shaggy on the body, and shorter and paler around the face. They are smaller than Alexander's cusimanses, and have a dark line that runs from the back of the neck to the base of the tail. Ansorge's cusimanses have a fairly long, snout-like muzzle, rounded ears, short legs, and a short, tapering tail. The feet have five digits, with quite long, sharp claws.

Ansorge's cusimanses live in deciduous rainforests in Central Africa (Democratic Republic of the Congo and Angola), where they partly overlap the range of Alexander's cusimanse. This is a social mongoose, with as many as 20 animals in a group. Ansorge's cusimanses are active during the day, and feed on insects, larvae, eggs, and small vertebrates.

There is very little information about this mongoose – up until 1984, only two museum specimens were known. Since Ansorge's cusimanses live in forests, they are likely threatened by the destruction of their habitat. They are hunted for bushmeat markets,

Ansorge's cusimanse (© Luke Hunter and Priscilla Barrett, 2018, A Field Guide to Carnivores of the World. Second edition, Bloomsbury Wildlife (UK), an imprint of Bloomsbury Publishing plc)

which undoubtedly reduces their local numbers, but it is not known if the current level of hunting pressure threatens their continued existence in some areas. The IUCN considers Ansorge's cusimanse to be a rare species, and acknowledges that habitat destruction and bushmeat hunting have probably impacted populations in the wild, but it does not believe that the declines in numbers are currently high enough to threaten the survival of this species.

Common cusimanse

Crossarchus obscurus

Also known as long-nosed cusimanse or West African cusimanse

Total length: 45–58 cm
Weight: 0.5–1 kg
Red List status: Least Concern

Common cusimanses have grizzled dark brown to blackish fur that is shorter and paler around the face than on the body. They have a long, snout-like muzzle, with short, rounded ears that can be closed by moving the outer ear ridges. Their short legs have feet with five digits (but the first are very reduced on the hind feet), and quite long claws. The tail is short and tapering.

Common cusimanses live in the dense understorey of West African rainforests and riparian forests, from sea level up to at least 1,500 metres; they have also been found in farm bush, logged forests, and plantations. Common cusimanses are social animals, with up to 24 individuals living together in a group. Each group contains one or more family units that

Common cusimanse (© Chris Stuart and Mathilde Stuart)

comprise an adult pair and their young from one or more litters. Common cusimanses are mainly active during the day, and will rest and sleep together in hollow logs, amongst dense vegetation, above ground in trees, or in burrows that are made by other animals. There is no permanent den site, and a group will often choose a new shelter every night.

Common cusimanses either forage on their own or together as a group, and will look for food within rotting logs and amongst the dense vegetation and leaf litter on the forest floor, using their long snout and strong claws to scratch, dig, and root out prey. They may also climb trees and probe in shallow water to find food. The most common food items are millipedes, ants, earthworms, and termites, but the diet also includes other invertebrates (snails, earthworms, spiders, crabs, woodlice, centipedes, grasshoppers, cockroaches and beetles) and small vertebrates (frogs, snakes, lizards, birds, eggs and rodents) as well as fruits and berries. Common cusimanses often look for food alongside Diana monkeys and sooty mangabeys, using them as sentinels for daytime predators, such as eagles. In response to the monkeys' alarm calls, the cusimanses will run for cover amongst dense vegetation or within hollow logs. Sometimes a cusimanse group will cooperate to hunt large rats and snakes: one member of a party goes down a hole to flush out a rat, which is pounced upon by others as it emerges. When faced with a cobra, the whole group will surround the snake and kill it. Small vertebrate prey are killed with a single bite at the back of neck, and invertebrate prey is shaken before being eaten. Millipedes emit a noxious substance when attacked that can irritate sinuses and eyes, and to combat this, common cusimanses stamp on the millipedes with their front feet and rub them in the soil until the irritating substance has dissipated.

A group of common cusimanses wanders throughout its territory and is rarely seen for more than a few days in the same place. The size of the territory depends on the group size and the quality of the forest, and can be up to 1.4 km^2. As the common cusimanse is a social species, individuals communicate with each other using a wide range of calls and sounds, including: a barking alarm call when confronted with a predator; a friendly greeting grunt when two individuals meet; and a whistling noise that is used for keeping in contact with each other when amongst thick vegetation, and which becomes louder and more frequent if an individual becomes separated from the rest of the group. One particular call is often heard when a group begins moving again after a long rest, and this is initiated by the leading individual.

Both males and females scent-mark objects using their cheek and anal gland secretions, which have a substantial difference in chemical composition between individuals. Common cusimanses do a reverse handstand, or lift their hind leg, to smear the anal gland secretions on vertical objects, and drag their anal region along horizontal surfaces. Group members will also use these secretions to mark each other, and will often deposit faeces and anal gland secretions at the same time, in the same area, especially near den sites and territorial boundaries.

Occasionally there are aggressive conflicts between individuals in a group, especially over food items, with the aggressor growling and baring teeth at its neighbour. If fighting does break out, this tends to be ritualised, and any biting is directed towards the pale cheek patches. An aggressive encounter was once seen between a group of four Liberian mongooses and a larger group of ten common cusimanses, which ended with the Liberian mongooses being the victors.

Although female common cusimanses can have several fertile periods throughout the year (polyoestrus), breeding in the wild appears to be seasonal, and births are most commonly seen during January, February, May, and June. Ovulation (the release of eggs) is induced by mating. The male initiates copulation by first mounting the female with his forelegs forward of her pelvic region. He then grasps the female at the back of the neck and begins thrusting. The gestation or pregnancy period is around eight weeks, and between two and five helpless pups, covered only with under-fur and with their eyes closed, are born in a den. The pups' eyes will not open until they are 12 days old. While pups are in the birthing den, a few members of the group, particularly the breeding pair, will guard and care for the young, while others leave for short foraging trips. After a couple of weeks, the pups will leave the den and may then be carried from one new den to the next, as the group moves throughout the territory. Weaning starts when the pups are three to four weeks old, and adults will then provide young with solid food for the next month or so. Young cusimanses are sexually mature at approximately nine months.

Crowned hawk-eagles are known to be a major predator of common cusimanses, and other large carnivores, including leopards, may also be a threat. A common cusimanse may live up to 13 years in captivity.

The clearance of forests for agriculture is undoubtedly having a serious impact on common cusimanse populations in the wild. And even though this mongoose species has been found in disturbed forests, it is not known how well it can tolerate logging activities or other habitat disturbances. The common cusimanse is heavily hunted with dogs and snares, which reduces local numbers, but no one knows by how much. The IUCN does not consider the common cusimanse to be endangered, as it believes this species is common and relatively widespread, and that the levels of habitat loss and hunting pressures are not currently severe enough to classify this species as threatened.

FLAT-HEADED CUSIMANSE

Crossarchus platycephalus

Also known as Cameroon cusimanse

Total length: 45–58 cm
Weight: 0.5–1 kg
Red List status: Least Concern

Flat-headed cusimanses have dark brown to black fur that is short and paler around the face, and is thick and shaggy on the body. They have a long muzzle, small rounded ears, short legs, and a short

Flat-headed cusimanse (© Organisation pour le Développement Durable et la Biodiversité (ODDB-ONG)

tapering tail. The feet have five digits, with quite long, sharp claws. Despite its common name, the flat-headed cusimanse has a broader, rather than flatter, head than the common cusimanse.

Flat-headed cusimanses are found in parts of West and Central Africa, and live in the undergrowth of dense rainforests, as well as in forest-savannah mosaics and forested farmland, from sea level up to 1,600 metres. This is a social species, with as many as 25 individuals in a group, although five to eight is more usual. Flat-headed cusimanses are mainly active during the day, and sleep together at night under fallen logs, inside hollow trees, amongst thick vegetation, or in burrows. They feed on forest floor invertebrates, by scratching and digging in the soil and leaf litter with their elongated muzzles and sharp claws, and also eat land and river crabs, small vertebrates, fruit, and berries. Individuals kill prey with a single well-directed bite and then crunch it up rapidly. Flat-headed cusimanses can climb sloping trees, and some have been seen pursuing a black cobra up in a tree, some 3 metres above the ground. They will also wade into shallow water, turning over rocks with their long muzzle in search of river crabs, which are grabbed with the mouth, tossed to the river bank, and then eaten. Flat-headed cusimanses might breed throughout the year, and the litter size may range from three to five.

The IUCN considers that the flat-headed cusimanse is not common anywhere in its range, but believes it to be a relatively widespread species that can adapt to habitat disturbances, and so may not be currently endangered. However, flat-headed cusimanses are very likely threatened by the clearance of forests for agriculture, and they are hunted for bushmeat by local villagers.

GENUS CYNICTIS

There is only one species in this genus, the yellow mongoose, which is found in southern Africa.

YELLOW MONGOOSE

Cynictis penicillata

Total length: 43–75 cm
Weight: 0.5–0.9 kg
Red List status: Least Concern

Yellow mongooses are reddish-brown to greyish-yellow, with a creamy-white chin, throat and upper chest, and a long bushy tail that can end in a pure to dirty-white tip. The ears are relatively large and project above the line of the head. There are five digits on the forefeet, with the first one reduced, and four on the hind feet. Northern yellow mongooses are smaller, more greyish, and lack a white tip to their tail.

Yellow mongooses are found across southern Africa, and live in a variety of open, semi-arid habitats, from sparse bushland to grasslands and semi-deserts, especially where there is soft or sandy soils for digging dens; they avoid woodlands, deserts, and mountains.

The social organisation of the yellow mongoose is somewhat intermediate between the solitary and social mongoose species. Although they are solitary foragers, several yellow mongooses will spend the night together in a communal den, and may cooperate

Yellow mongoose (South Africa © Emmanuel Do Linh San)

in raising young. The size of a group can be up to 13, but is usually only three or four, and each group typically has an established hierarchy of one dominant male, subordinate females and males, and juveniles. There is some geographical variation in the social structure: in some areas, groups share a communal territory that is defended by both males and females, whereas in other places, adults of both sexes occupy different areas of a large, loosely-shared communal range, in which males can have home ranges around 1 km^2 in size that substantially overlap each other and those of several females, while female home ranges are generally smaller, up to 0.5 km^2, and quite separate, except at the communal den site. Both male and female yellow mongooses move similar distances whilst looking for food (around 250 metres per hour), and males can cover several kilometres in a day.

Yellow mongooses are terrestrial and mainly diurnal, but can also be active at night, especially when there is exceptional food availability, such as termites swarming. Several individuals sleep together in burrows, which they enter around sunset and exit shortly after sunrise; they may also occasionally rest around midday for variable periods of time. Burrows may be quite simple, with only one or two entrances, or they are much more complex, having as many as 60 openings, with several linked underground tunnels and nest chambers at different levels, and up to 1.5 metres in depth. Yellow mongooses can dig their own burrows, or modify existing burrows, and sometimes they share them with meerkats or ground squirrels. Yellow mongooses and meerkats that live together will cooperate in looking out for predators, which increases their chances of detecting dangerous animals. Yellow mongooses are partly nomadic, and may rotate the use of several burrow systems on different nights.

Yellow mongooses predominantly eat insects, preferring termites, beetles, locusts, and grasshoppers. However, they are opportunistic feeders and will also hunt rodents, birds, reptiles, amphibians and arachnids, such as scorpions and spiders; sometimes they will eat fruits and carrion, and occasionally raid hens and their eggs. Their diet varies both geographically and seasonally. Yellow mongooses look for food on their own, but are sometimes seen in pairs or in small groups, and will typically forage during the early morning and late afternoon.

When two yellow mongooses in a group meet, they sniff each other's facial glands to assess their dominance status. The more dominant individual will rise higher on its feet and bite the subordinate's neck. The subordinate will then lie on its side and may emit a high-pitched scream. A dominant individual will also mark a subordinate by first straddling it from above and then use its anal glands in a standing position. In some yellow mongoose colonies, the dominant male marks other members on a daily basis, whereas in others, the dominant male may not be involved in dominance interactions, but high-ranking subordinates still defer to the dominant male, and in turn dominate lower ranking members.

Yellow mongooses are usually quite quiet, but five vocalisations have been heard: a high-pitched scream that is uttered during fighting; a low growl when an individual is bothered at a food source; an even lower growl when an individual is stressed; a short barking sound that is emitted as an alarm call; and a soft purring that is uttered during copulation. Yellow mongooses use the secretion from their anal glands to scent mark home ranges and other members of their colony. The cheek glands may also be used to mark objects, and this is often preceded by wiping the entire side of the body on the ground or side-swiping. All members of a group help scent mark, although most marking is done by younger, subordinate individuals. In yellow mongoose groups that defend a communal territory, middens or latrine sites are located along the boundaries, whereas in those colonies in which each adult has its own home range, droppings are deposited in a depression or hollow close to burrow entrances.

Yellow mongooses breed during the wet summer, from August to February, with two litters produced in quick succession: one around early to mid-October and another during December to February. There is no reproductive suppression by the dominant pair, so more than one female in a group can breed at the same time. Mating starts in early July, and when females are in oestrus, males purr, caw and scream while following the females around and attempting to mount. Oestrous females will allow copulation over a two-day period, after which males are vocally rebuffed, with bites to the head and neck. During copulations, which can last up to 45 minutes, the male purrs while the female bites or licks the male's ears and neck. The gestation period is 60–62 days, and up to five pups are born in a litter, although two is more common. The pups are born within a nesting chamber in the burrow, and are given milk for the first 6–8 weeks. In some groups, individuals may guard, feed and groom pups that are not their own. Any young from the previous year will help feed pups at the den by carrying back large prey items, such as rodents, reptiles, and large spiders. Pups first accompany adults on foraging expeditions at about eight weeks of age, and are fully independent when they are around 16–18 weeks old.

Large snakes and monitor lizards prey on young yellow mongooses, and large raptors, such as tawny eagles and martial eagles are also important predators. In captivity, yellow mongooses can live up to 15 years.

Yellow mongooses live in quite high densities across southern Africa and may be tolerant of human-induced changes to their habitat, so their long-term survival in the wild looks positive. However, they are a carrier of rabies, which has led to several extermination campaigns in some areas, usually through the use of burrow gassing. Yellow mongooses were also persecuted in the past because it was believed that they were a predator of newborn lambs, but this myth has now been totally discredited.

Genus *Dologale*

There is only one species in this genus, Pousargues' mongoose, which is found in Central Africa. Pousargues' mongoose is known from just 31 museum specimens, a handful of possible sightings and a few recent photographs and observations, so we know almost nothing about this species. It is one of the smallest mongooses, and bears a close resemblance to the common dwarf mongoose.

Pousargues' mongoose

Dologale dybowskii

Also known as savannah mongoose

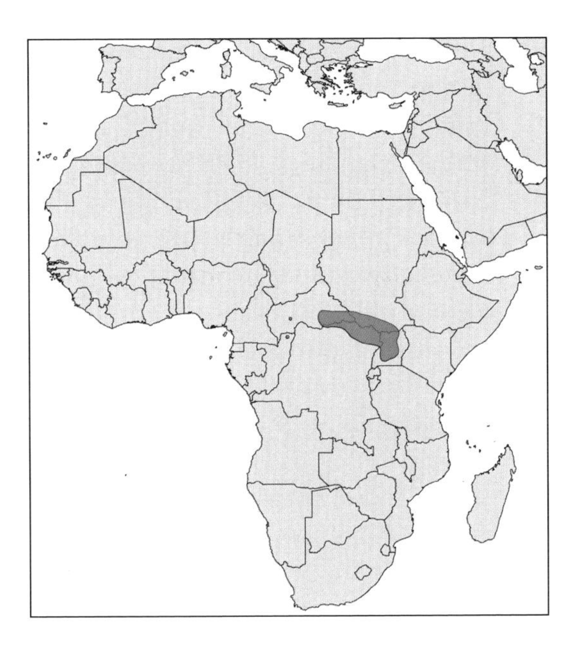

Total length: 41–56 cm
Weight: 0.3–0.4 kg
Red List status: Data Deficient

Pousargues' mongooses are grizzled dark brown, with a blackish-grey head and neck, and underparts that are reddish or pale grey. They have a relatively short muzzle, a prominent ruff or 'cowlick' of fur on the neck, dark-coloured short legs that have almost black feet, and a quite long bushy tail. The feet have five digits.

Pousargues' mongooses are found within a small region of Central Africa, and live in savannah-forest mosaics, montane forest grasslands, and the thicketed shores of Lake Albert. In the southeastern part of the Central African Republic, a population of Pousargues' mongooses was recently discovered in the Chinko/Mbari drainage basin, where eight other mongoose species have also been recorded.

Pousargues' mongoose (Uganda © Jason Woolgar)

It appears that Pousargues' mongooses are at least partly diurnal, and they may rest in holes in trees and termite mounds. Recent observations have revealed that they live in small groups, ranging from 3–12 individuals, which move between termite mounds on a regular basis. Their strong digging claws and unspecialised teeth suggest that their diet may include invertebrates and small vertebrates, and one museum specimen had millipedes, termites, and small seeds in its stomach. They pounce on insects on the ground, and frequently dig in loose soil and turn over light stones to look for food. Although individuals may largely feed alone, they keep in contact with each other using a series of vocalisations reminiscent of other social mongoose species. Breeding was seen taking place in mid-May during the early rainy season, and a litter can be up to four pups. A small group of three adults was seen caring for an infant, carrying it around in their mouths.

We do not know the exact range of this mongoose, its habitat preferences, the status of populations in the wild, and what possible conservation threats that this species might face, which is why it is currently classified as Data Deficient on the IUCN's Red List, as there is insufficient information to make a proper evaluation of its conservation status and extinction risk. Clearly, we urgently need field studies on this species in order to gather the necessary distribution, ecological and conservation data for a more informed conservation assessment.

Genus *Galerella*

The *Galerella* mongooses, which all occur in Africa, have previously been included in the genus *Herpestes*, and in some publications they are still listed as *Herpestes* species, but recent morphological and genetic analyses strongly suggested that they should be included in their own genus, *Galerella*, which we have followed for this book. In addition to this disagreement over genera, there is also a debate amongst taxonomists over how many *Galerella* species there are, with as many as five proposed by some authors. Based on morphological features, four *Galerella* mongooses have been long established and generally accepted as valid species: Kaokoveld slender mongoose, Somali slender mongoose, Cape grey mongoose, and common slender mongoose. These last two mongooses have been included in recent genetic studies, which confirmed that they are distinct species.

A *Galerella* subspecies, which is found in Namibia and Angola, has been advocated as a fifth *Galerella* species, the black mongoose (*Galerella nigrata*). This mongoose is blackish in colour and lives in mountainous areas that have large granite boulders and rocky outcrops, known as inselbergs. In the past, there was much confusion as to which *Galerella* species this subspecies belonged to, and even though a recent molecular study has provided some genetic information on this population, this new evidence is not yet fully conclusive and further studies are still needed to confirm its status. At the moment, it is very difficult to know how this particular mongoose entity should be treated, but for the purposes of this book, we have decided to include it with the Kaokoveld slender mongoose.

Kaokoveld slender mongoose

Galerella flavescens

Also known as Angolan slender mongoose (and includes the black mongoose)

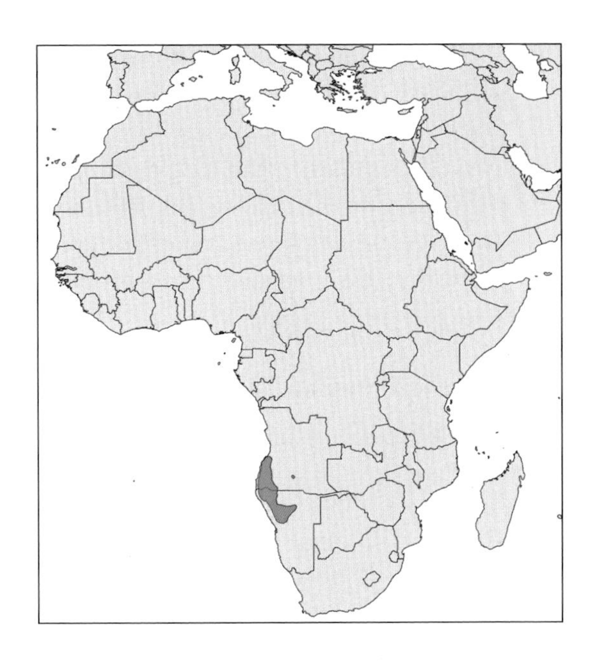

Total length: 64–72 cm
Weight: 0.5–0.9 kg
Red List status: Least Concern

Kaokoveld slender mongooses have a long, slim body and tail, and are reddish (or dark brown to black, for the subspecies *nigrata*). Non-black individuals have a tail with a black tip and underparts that are buff-coloured.

Kaokoveld slender mongooses are found in southern Africa (Angola and Namibia) and live in arid habitats that provide cover, and in woodlands dominated by large granite boulders (subspecies *nigrata*). They are active during the day, and are usually solitary, occupying home ranges that can be up to 1.5 km^2. Kaokoveld slender mongooses are opportunistic foragers and their diet includes insects, scorpions, small mammals, birds, lizards and snakes, as well as freshwater crabs and other aquatic animals. Up to seven Kaokoveld slender mongooses were once seen feeding on the larvae and flies covering a dead, rotting antelope carcass. While one animal fed, the others would wait nearby, and would only move in to feed when no others were present. On five occasions, short duration fights broke out between individuals – initial growling was followed by face-to-face biting, rapid rolling around on the ground, and loud screeching or squealing.

Kaokoveld slender mongoose (Namibia © Hans Hillewaert, CC BY-SA 3.0, Wikimedia commons)

The Kaokoveld slender mongoose has a relatively restricted distribution, but the IUCN believes it to be common and facing no known major threats, and so considers this species not to be endangered. Encroachment of local communities, with dogs and livestock, has led to the disappearance of Kaokoveld slender mongooses from several areas in northern Namibia, possibly due to dog predation and the trapping of mongooses to reduce chicken losses. Careful monitoring of populations are thus needed to ensure that this species continues to survive across its range.

Somali slender mongoose

Galerella ochracea

Also known as Somalian slender mongoose

Total length: 47–56 cm
Weight: 0.3–0.8 kg
Red List status: Least Concern

Somali slender mongooses have a long, slim body and tail, and are dark brown to pale grey, sometimes reddish. They are found in East Africa (Somalia, Ethiopia, and Kenya) and live in semi-arid, open woodland and hilly areas, up to at least 600 metres. They appear to be diurnal, and are thought to be solitary.

This is a very poorly known mongoose and there is virtually no information about its natural history, habitat requirements, status of populations in the wild, or the impact of any potential conservation threats. However, the IUCN assumes that Somali slender mongoose is locally common and does not face any major threats, and so considers it to be not endangered. This is one species for which field studies are very much needed.

Somali slender mongoose (© Luke Hunter and Priscilla Barrett, 2018, A Field Guide to Carnivores of the World. Second edition, Bloomsbury Wildlife (UK), an imprint of Bloomsbury Publishing plc)

Cape grey mongoose

Galerella pulverulenta

Also known as small grey mongoose

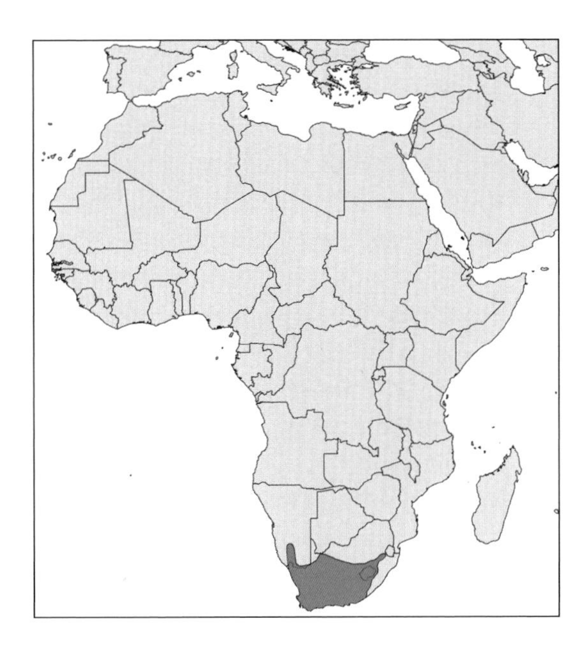

Total length: 54–76 cm
Weight: 0.5–1.3 kg
Red List status: Least Concern

Cape grey mongooses have a grizzled light to dark-grey coat, with slightly darker muzzle and legs, and a long, bushy tail. Their ears are small and close to the head, and there are five digits on each foot, with the first one reduced.

Cape grey mongooses are found across southern Africa, from sea level up to 1,900 metres. They live in a wide variety of habitats, from semi-desert to high-rainfall coastal forest, especially in areas that provide cover. Cape grey mongooses avoid open terrain, but are sometimes found close to human settlements. They seem to be more common in dry than in wet areas, although they are absent from the driest parts of the region.

Cape grey mongooses are diurnal. They are particularly active around sunrise and sunset, less so in bad weather and at high temperatures, and will rest in termite mounds, within dense vegetation, under rocks, or in burrows. They are poor diggers and do not dig their own burrows, instead relying on holes dug by other species. Although Cape grey mongooses are mainly terrestrial, they can climb trees. They are usually solitary, but family groups of up to five are sometimes seen. Home ranges can be up to 0.9 km^2, and these can overlap considerably between individuals. Adult males may occasionally associate loosely in pairs.

Cape grey mongoose (© Chris Stuart and Mathilde Stuart)

Cape grey mongooses are opportunistic predators and their diet includes a wide range of prey, from termites and grasshoppers to snakes and rodents. Nevertheless, their staple foods are generally small mammals and, to a lesser extent, insects. They will also feed on larger mammals, such as hares and porcupines, but as Cape grey mongooses do scavenge, it is uncertain to what extent they can catch large prey. They usually move from bush to bush while foraging, inspecting holes and rodent nests, and relying mainly on sight and smell for finding food. Cape grey mongooses will scratch the soil in search of invertebrates, but they are not avid diggers. Insects caught on the ground are held down with the front feet and then eaten. Small mammals are stalked and then killed with a bite delivered to the head. Larger prey is held firmly on the ground with the front feet and then torn apart. Cape grey mongooses can break eggs by throwing them against a vertical surface. While foraging, individuals may move 50–100 metres every 15 minutes, and travel on average 4 kilometres during the course of a day.

Cape grey mongooses scent mark using the secretion from their anal glands. Unlike many other mongooses, they do not leave droppings in faecal piles, but scatter them singly in well-used areas, especially around sleeping sites. Because of their small size, they are at high risk of being killed by a variety of predators, including martial eagles, caracals, and marsh mongooses.

Cape grey mongooses are seasonal breeders, particularly from August to December. The gestation period is approximately 60 days, and one to three pups are born and raised in rock crevices, farm outbuildings, fodder stores, and woodpiles.

The IUCN considers that the Cape grey mongoose is not endangered, as this species is believed to be common and adaptable, it is present in a number of protected areas within its range, and no major conservation threats are known.

COMMON SLENDER MONGOOSE

Galerella sanguinea

Also known as slender mongoose

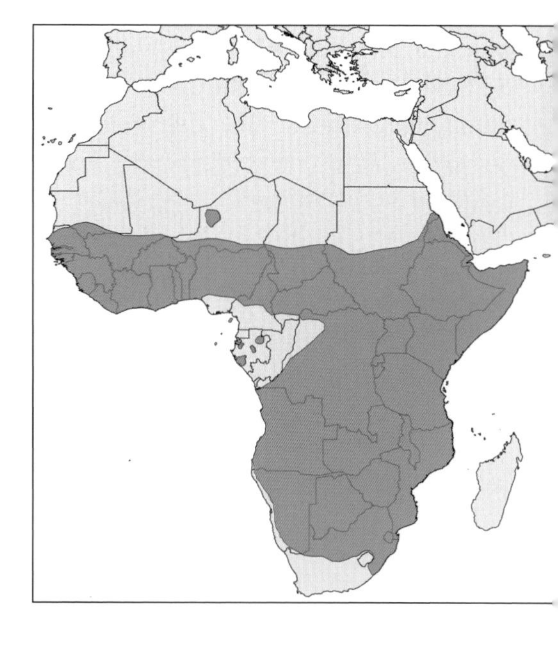

Total length: 60–96 cm
Weight: 0.4–0.8 kg
Red List status: Least Concern

Common slender mongooses have a lithe body and a long, slim tail, which has a distinctive black tip. Their fur can vary widely in colour, from dark reddish-brown to grey, or even yellow, and has a grizzled appearance due to the lighter-coloured bands on each hair. The face is pointed, and the ears are

Common slender mongoose (Botswana © Laurent Vallotton)

small and close to the head. The nose is small, and usually pinkish-brown to light brown. Each foot has five digits, with the first one smaller than the others.

Common slender mongooses are found throughout most of Africa south of the Sahara, from sea level up to 2,700 metres. They live in a wide variety of habitats within arid and semi-arid areas, and even close to villages, but are absent from true deserts and dense rainforests, although they may sometimes skirt forest fringes or penetrate into forests along roads.

Common slender mongooses are diurnal, with a slight peak in activity before sunset. They rest and sleep in termite mounds, rock piles, under tree roots, at the base of trees, in hollow logs, or in burrows that they have dug themselves or usurped from other animals. Burrows often have a single entrance that are connected by short passages to large chambers. Although largely terrestrial, common slender mongooses can climb, and although not great climbers, they have been seen running up and down wire netting, stone walls, trees, and branches.

Common slender mongooses are mainly solitary. Males have home ranges of around 0.5-1 km^2 that do not overlap each other, but often encompass the ranges of several females, which have smaller, non-overlapping ranges of around 0.25 km^2 in size. Two to four closely-related males sometimes form a loose, non-aggressive association that jointly defends a territory, which may overlap the home ranges of up to six females.

The diet of common slender mongooses includes a wide variety of small vertebrates and invertebrates, and some wild fruits. Insects, such as grasshoppers, termites, beetles and ants are found very frequently in the stomach of specimens and in faeces, but small vertebrates, including rodents and lizards, are often more important in terms of the total amount consumed. Since common slender mongooses are better climbers than many other mongoose species, they are able to catch and eat birds. They will also feed on carrion, along with any infesting insects. Their diet seems to follow seasonal variation in availability: insects are more frequently consumed during wet and warm periods than in dry and cold ones, when more small mammals and reptiles are eaten. While hunting, common slender mongooses move around continuously, chasing and pouncing on rodents, and catching in flight any flying insects that they flush. They move swiftly and silently, sniffing and searching visually, without pausing to scrape and scratch at crevices. Captive common slender mongooses are fond of eggs, which they break open by smashing them against a hard surface.

Common slender mongooses are often seen travelling along roads or pathways, with their nose to the ground, an arched back, and the tail held low, with only the dark tip raised. When disturbed, they may stand motionless, and if pursued, the tail flips up into a vertical position. Five different vocalisations have been heard: a loud, sharp spit and a growl, when threatening a conspecific or warding off attackers; a buzz, which is used in agonistic encounters; a snarl, when two animals approach each other in attack; and a 'huh-new' distress call. Captive common slender mongooses will also threaten each other by snapping and spitting, and less dominant individuals will turn their lowered head away and retract their lips to show the pink gums, or approach in a submissive posture. In the wild, both males and females make use of latrines for urination and defecation, and during the mating season, males will scent mark with the secretion from their anal glands.

Common slender mongooses are seasonal breeders, and the birth of pups is concentrated in the wet season, which lasts from October to November and February to April in East Africa, and October to March in southern Africa. Females may have two litters per year. While a female is in oestrus, which may last more than a week, the male will follow her closely, and will attempt to mount whenever she stands still. Copulations last up to three minutes. The gestation period is around 60–70 days, and between two and four pups are born in a den, which is then changed every three to four days, with the pups being carried from one den to the next. The pups' eyes are closed at birth, but are fully open by the time they are three weeks old. Pups begin eating solid food at four weeks, and are fully weaned at around seven to nine weeks. They grow rapidly, reaching two-thirds the weight of an adult in 50 days, and full adult size and sexual maturity by the age of one year. Juveniles, particularly males, disperse before they are six months old.

The main predators include raptors, such as the African hawk-eagle, tawny eagle and martial eagle, and larger carnivores, such as the lion, leopard and African wild dog, may also take individuals. The oldest common slender mongooses found in the wild were eight years old, and a captive animal lived to nearly 13 years.

As their name implies, common slender mongooses appear to be quite common across most of Africa, and they occur in numerous protected areas. No significant human-caused threats are known, although common slender mongooses are captured for bushmeat markets and for use in traditional medicines.

GENUS *HELOGALE*

There are two very small mongooses in this genus, the Somali dwarf mongoose and common dwarf mongoose, which are both found in Africa. The common dwarf mongoose is the smallest mongoose in the world, and is one of the better known and more intensely studied species. The ranges of these two closely-related mongooses overlap in the horn of Africa, but it not known how they can coexist in the same region, although the Somali dwarf mongoose seems more adapted to arid conditions.

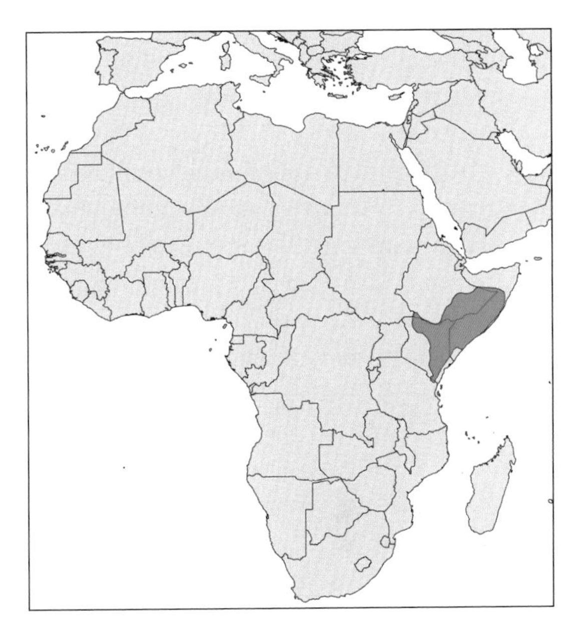

SOMALI DWARF MONGOOSE

Helogale hirtula

Also known as Ethiopian dwarf mongoose, desert dwarf mongoose

Total length: 35–45 cm
Weight: 0.2–0.4 kg
Red List status: Least Concern

Somali dwarf mongooses are a grizzled grey colour, with a yellowish face and underparts. They have a slender, elongated body, short legs, dark-coloured feet, and a tapering tail. The pointed muzzle is quite short, and the ears are small and rounded. Each foot has five digits, with the first one reduced.

Somali dwarf mongoose (© ZooParc de Beauval)

Somali dwarf mongooses are found in the Horn of Africa region (Somalia, Ethiopia and Kenya), and live in arid to semi-arid open woodlands, scrub and grassland, up to 600 metres. Somali dwarf mongooses are active during the day, and rest and sleep in termite mounds and amongst rocky outcrops. This is a social species, but we know nothing about its social organisation or breeding behaviour in the wild. Due to their small body size, preference for open habitat and diurnal lifestyle, it is likely that Somali dwarf mongooses are vulnerable to a wide array of daytime predators.

The Somali dwarf mongoose has a patchy distribution throughout its range, and as virtually nothing is known about the status of populations in the wild and which threats this species might face, assessing its conservation status is very difficult. The IUCN has currently classified this species as Least Concern, as it is presumed to occur in several protected areas across its range and may not face any major threats, but this conservation assessment could change once more information about this mongoose comes to light.

Common dwarf mongoose

Helogale parvula

Also known as dwarf mongoose

Total length: 31–41 cm
Weight: 0.2–0.3 kg
Red List status: Least Concern

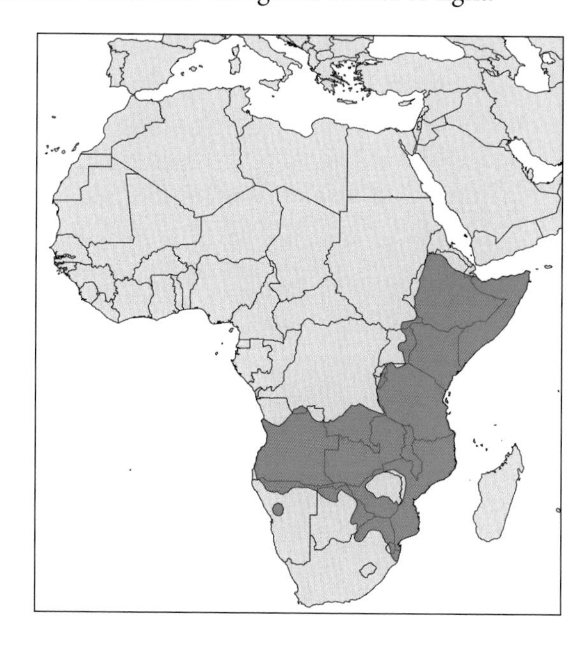

The coat colour of common dwarf mongooses varies from yellow to dark brown and has a grizzled appearance, due to lighter-coloured bands on each hair. They have a domed head, with a pointed muzzle, and small, rounded ears. Their feet have five digits, with the first one reduced, and strong claws.

Common dwarf mongooses are found along eastern Africa and also westwards to Angola and Namibia. They live in open woodlands, thickets, and wooded savannahs, particularly where there is a high density of termite mounds or other suitable den sites. Although they can occur in semi-desert regions, common dwarf mongooses are not found in very arid areas. They live in groups of up to 32 individuals, although around nine is most common, with each group comprising both related and unrelated adult males and females, and young from one or more litters. Each group defends a territory that can be up to 1km^2 in size. There is little overlap between territories, but when two rival groups do meet, the smaller one will generally retreat without any physical aggression, although skirmishes can sometimes occur.

Common dwarf mongoose (Zimbabwe © Emmanuel Do Linh San)

Common dwarf mongoose groups are highly social, with each member cooperating in defending a shared territory, watching for predators and rearing young, as well as babysitting pups and grooming each other – they will even provide care for a sick or injured pack member. Within each group, there is dominance hierarchy, with a single dominant breeding pair, usually the oldest male and female, and subordinates of each sex that are reproductively suppressed but help to raise the young of the alpha pair. The mongooses that achieve alpha status bond for life, and can have a tenure of several years. When one of the alpha pair dies, it is replaced from within the pack by the oldest adult of the same sex. The alpha female presides over all other group members in access to food, and can displace others from a foraging site. The male dominance hierarchy is most obvious during the breeding season, when males compete for access to females.

Common dwarf mongooses are active during the day, and they sleep together overnight in dens, especially in termite mounds or amongst rock crevices, emerging in the early morning and returning to the den before sunset. There may be up to 200 termite mounds within a territory, and common dwarf mongoose groups favour areas that have the highest number of them. Each group generally occupies a different den every night, but will use a particular one for longer periods when raising young.

Common dwarf mongooses feed almost entirely on insects, especially termites and beetles, but they will also eat other invertebrates, such as millipedes and scorpions, and occasionally small mammals, lizards, snakes and birds, as well as some fruits. They catch flying insects in the air with their forepaws and pounce on any insects that are moving along the ground, and will also dig in the soil and leaf litter for invertebrates. Large insects are pinned down with a foot and then bitten on the head, and are usually eaten head first. Small vertebrates, such as mice, are also pinned down with the forefeet, then killed with a bite through the back of the head. A small snake is bitten just behind the head, dropped, then repeatedly bitten and shaken violently until it can be pinned down with the forepaws and eaten, starting at the head. Larger snakes are attacked by the whole group, and may be shared by everyone. Common dwarf mongooses seem quite nonchalant about eating stinging prey, such as scorpions, despite being stung. Small bird eggs are bitten open at one end and licked out, whereas a larger egg is grasped by the forelimbs and then thrown backwards through the hind limbs to crack open on a rock or other hard surface.

In the early morning, common dwarf mongooses start the day by sunning themselves near the entrances of their den site, and also spend time grooming both themselves and others. Grooming pairs are usually of the same sex, and are typically similar in social rank within the pack; most adults will also groom youngsters. After this early activity, the alpha female then initiates pack movements, but before moving out, babysitters are left to look after any pups that are too young to accompany the group on the foraging trip. During the rest of the day, the remaining pack members forage together, often resting around midday. Daily foraging distances can be up to 1 km, with larger groups travelling further than smaller ones, and shorter distances being covered when travelling with young. As the group moves slowly as a loose unit, pack members spread out over an area roughly 10–200 square metres, within which they look for food on their own, whilst keeping in close vocal contact with each other. Guards or sentinels, especially subordinate males, will watch for predators while the others forage, and move to a vantage point, such as a termite mound, from which to scan the area. If the sentinel sounds an alarm call, the group will freeze, or head straight for the nearest cover if the danger proves immediate. In the absence of any danger, the sentinel will return to foraging after a few minutes, and another group member will take over the guard duty.

Common dwarf mongooses communicate with each other using a variety of vocal sounds: short, nasal 'peeps' are emitted every few seconds to maintain group cohesion while foraging; a 'moving out' call is given by an alpha female to entice the rest of the group to follow her; shrill war cries alert group members to the presence of a rival group (pack members will then bunch and charge, with the alpha male leading and the alpha female at the rear); repeated alarm calls warn the group of a predator, and an alarm 'chitter' is given as they run to cover. Different alarm calls indicate whether the predator is approaching from the sky or along the ground, and these calls are shorter and higher in pitch, the more immediate the danger.

Common dwarf mongooses will also respond to the alarms calls that hornbills give out when a raptor approaches. Early in the morning, hornbills will perch upon the termite mound in which the group of mongooses has taken refuge overnight, and wait for them to rouse. As the mongooses forage through the shrubs and grass, the hornbills snap up any insects that are disturbed. If a hornbill spots an approaching raptor, it will utter an alarm call, sending the mongooses scurrying to the nearest shelter. Should the mongooses happen to sight a raptor first and give out an alarm call, the hornbill will also flee. Hornbills will utter an alarm call for raptor species that do not prey on them but are predators of the mongooses. Common dwarf mongooses also known to form a similar relationship with other co-foraging bird species, such as fork-tailed drongos.

Both sexes use the secretions from their anal and cheek glands to scent mark each other. Common dwarf mongooses also deposit scents on various objects within the group's territory, especially near the termite mounds that they sleep in, by dragging their anal region over horizontal surfaces and using a reverse handstand against vertical structures. Interestingly, male anal gland secretions contain vitamin E, but this is absent from female secretions, which suggests they might also play a sex-specific role. Faeces and urine are deposited in communal latrines.

Common dwarf mongooses tend to breed during the wetter months of the year, when invertebrate prey is most readily available, and only the dominant pair usually have young, but this is not always the case. The fertile period (oestrus) of all the females in a group is synchronised, and lasts for one to seven days. Oestrus can occur within two to four weeks of giving birth to pups, which enables a female to have up to four litters per year. Early in the alpha female's oestrus cycle, the alpha male guards her and aggressively repels subordinate males, who attempt to sneak copulations. The alpha pair will consort all day, and may engage in as many as 50 copulations every hour, each one lasting from several seconds to 11 minutes. Later in the oestrus period, the alpha male will mate with subordinate females, and the alpha female may mate with other males. Subordinate females will also try to mate with males of all ranks, although the alpha male will aggressively interfere with the mating attempts by male subordinates. Despite the successful matings that female subordinates do manage to achieve, they often fail to become pregnant, due to a hormonal suppression that prevents the fertilization or implantation of eggs. However, some female subordinates, usually older ones, do become pregnant every year, and they may account for nearly a third of the pregnancies in a group. In fact, subordinate females may give birth to 15% of the offspring in a group, and subordinate males sire approximately 25% of the pups.

The gestation period is approximately seven weeks, and two to six pups are born in a den (usually a termite mound), with two or three being the most common. The birth of a subordinate female's pups is often synchronised with those of the alpha female, and may even happen on the same day. Most of these pups, however, are killed by the alpha female, but any that manage to escape this infanticide will be raised in a joint litter with the alpha female's pups. Even though some subordinate females have not given birth, they will nurse the alpha female's pups, and are able to provide milk due to having a pseudopregnancy. The weaning of pups occurs when they are around 40–45 days old.

The pups are kept in the subterranean den until they are approximately three to four weeks old. During this period, one or more individuals, usually subordinate females, will remain at the den to babysit the pups, while the rest of the group forages. These babysitters guard the communal litter from predators, as well as groom the pups. They remain at the den for several hours, until relieved by another set of babysitters. Litters may occasionally be left unguarded for a few hours, but usually at least one pack member, and sometimes as many as three or more, stands guard until the young are old enough to leave the den. When the pups do emerge from the den, and begin accompanying the foraging group, adults will provide them with food, and also carry young pups to new dens; subordinates, mainly females, usually provide more care than the alpha parents. Pups will beg for food from adults, who will bring food to them, dropping whole or partially eaten prey items.

Pups also learn to forage by following an adult, first observing, and then later being allowed to dig and capture prey items.

Both males and females will disperse from a group, usually as young adults and often as littermates of the same sex, but females are more likely to remain in their natal group than males. Dispersers can increase their likelihood of attaining a dominant breeding position by emigrating to another group.

Common dwarf mongooses are small animals that live in open habitats, and are active during the day, which makes them very vulnerable to a variety of daytime predators. Pups are particularly at risk from snakes, monitor lizards, raptors (including pale chanting goshawks and brown snake eagles), and also larger mongoose species, such as the banded mongoose and Egyptian mongoose. In the wild, common dwarf mongooses can live up to 14 years and, in captivity, one individual lived for 18 years.

The common dwarf mongoose is distributed across a large area of Africa, can occur at high densities within its range, is present in several protected areas, and currently faces no major threats, so it is unlikely that this mongoose species will be threatened in the foreseeable future.

GENUS *HERPESTES*

As many as 15 mongoose species have been included in this genus, but recent genetic studies have indicated that most of them should now be placed in other genera, such as *Atilax*, *Galerella*, *Urva* and *Xenogale*, in order that the systematics follows the evolutionary history of the group, as it should. But new taxonomic changes can take some time to appear in the literature, so in some publications many species are still listed under the genus *Herpestes*. We have decided to follow the new taxonomic proposals, which means that there is now only one mongoose species in this genus, the Egyptian mongoose, which is, in fact, the type species of *Herpestes*. This species is also called ichneumon, a name that is derived from the Ancient Greek word for 'tracker', and which likely reflects its ability to find and dig out crocodile eggs.

EGYPTIAN MONGOOSE

Herpestes ichneumon

Also known as large grey mongoose or ichneumon

Total length: 101–117 cm
Weight: 2–4 kg
Red List status: Least Concern

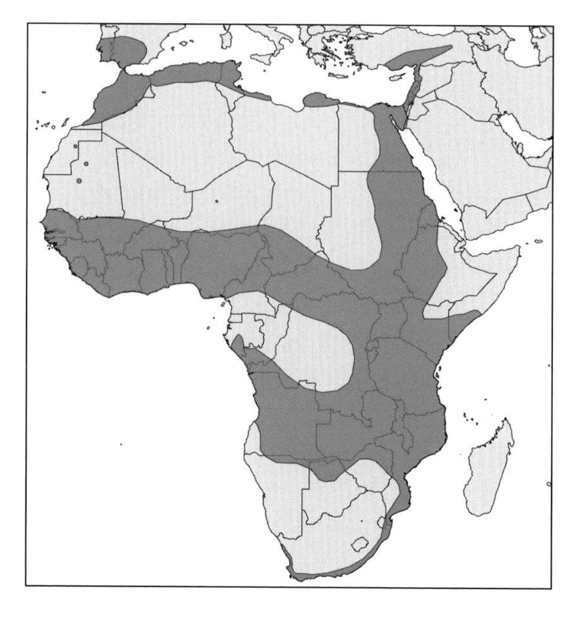

Egyptian mongooses are a grizzled grey, with a dark face, dark lower limbs and a black-tipped tail. They are long-bodied, with short legs, a pointed

Egyptian mongoose (© Chris Stuart and Mathilde Stuart)

muzzle, and short, rounded ears. Their feet have five digits, although the first one is quite short on the hind foot, and each has curved claws.

Egyptian mongooses are found across a large area of Africa, as well as southern Europe (Spain and Portugal) and parts of the Middle East. They live in a wide range of habitats, including forests, grasslands, wetlands, semi-deserts and montane areas, particularly in dense understorey vegetation and alongside rivers, lakes and swamps; they may also occur in cultivated areas and on farmland.

Egyptian mongooses are mostly solitary, but occasionally are seen in pairs and family groups of up to six individuals. In South Africa, the home range of males and females averages 0.4 km^2 in size; while in Spain, the home ranges are around 3 km^2, with males having exclusive home ranges that each overlap several females, whereas the home ranges of females overlap substantially with each other, except for their core areas. Although Egyptian mongooses are normally a solitary species, they can be flexible in their social organisation under certain circumstances. In Spain, groups of up to five individuals may form, comprising an adult male and female, together with their young, which will share resting sites and dens after spending the day alone. In Israel, several Egyptian mongooses formed territorial, social groups around a garbage dump, with four groups occupying a total range of 3 km^2.

Egyptian mongooses are mainly diurnal, but in some parts of their range may also be active during the night or twilight hours, around dawn and dusk. They rest and sleep in underground dens, amongst thickets, and within tree hollows. Egyptian mongooses are entirely terrestrial, but are also good swimmers, and can dig with their strong forelimbs and long claws.

Egyptian mongooses are opportunistic predators, feeding on the most abundant food available in each area and season, and eating a wide variety of invertebrates (including insects, slugs, and spiders), small vertebrates (rodents, rabbits, birds, lizards, frogs, snakes, crabs, and fish), some fruit and fungi, and carrion. They normally hunt alone, although in Spain, two or three individuals were seen together excavating rabbits from their underground dens. While foraging, Egyptian mongooses walk with their nose close to the ground and search intensely for prey in small holes in the ground and burrows, around bushes, shrubs and fallen trees, and will also dig frequently. Adults kill small prey with a bite to the head and larger prey with a bite to the neck, and will start eating from the head. Snakes are attacked from the rear before they can retaliate. Egyptian mongooses may break eggs open by throwing them backwards through their hind legs against a hard surface.

Egyptian mongooses use a variety of sounds, scents and visual markers to communicate with other mongooses. They have several, distinct vocalisations: a deep, sharp growling alarm call that elicits fleeing in other individuals; a short, repeated contact call given by family members during foraging; a growl associated with defence of food, territory or mate; a bark or spit given during mating or fighting; and a short, sharp, vigorous pain call. Egyptian mongooses will smear the secretion from their anal glands on stones along trails, either by squatting, or by dragging their anal region along an object – this secretion is composed of complex long-chain carboxylic acids, which have different compositions in males and females. Latrines are frequently found in open places in and around thickets, or outside underground dens, and scattered faeces are deposited in the centre of paths within the most used areas, and also on trails leading to resting sites.

Breeding is seasonal, with the birth of pups peaking during May to July in Spain, September to February in East Africa, and October to December in southern Africa. During the breeding season, males make more frequent contacts with females. Prior to mating, the male closely follows the female, sniffing and licking her vulva, and engaging in prolonged social grooming. During copulations, which may last up to seven minutes, the male embraces the female's midsection and thrusts with his hind legs pressed against her; afterwards, the female is restless and makes defensive threats towards her mate. Gestation lasts around 67 days, and up to four pups are born in a litter. Females usually have only one litter per year, but can produce a second litter if the first one is lost, or if the rodent population is high. The pups have closed eyes and ears at birth. Their eyes open by day 21, and the sense of smell is well-developed by day 39. Although pups can stand and walk when three weeks old, only in their fourth week can they walk properly. Weaning occurs between four to eight weeks, and pups first appear outside dens when they are about six weeks old. At around two to three months, the young have all their teeth except the molars, which appear about three months later. By the time they are four months old, young Egyptian mongooses can hunt by themselves. In Spain, females reared the pups alone, but in Israel, pups suckled from any breeding female in the communal group, and group members helped bring food to them. The young reach sexual maturity when they are over one year of age.

Egyptian mongooses are frequently killed by Iberian lynx in Spain, and large eagles are important predators throughout this species' range. The oldest known individual in captivity lived to 13 years.

Although the Egyptian mongoose has been relatively well-studied across the northern tip of its distribution, we have very little information about its natural history and any conservation threats over the largest part of its range in Africa. Its dependence upon riparian habitats would make this species vulnerable to any harmful impacts caused by water pollution and drainage. On the Iberian Peninsula, incidental and deliberate poisoning is a localised threat. In Portugal, stand hunting and hunting mongooses with dogs are legal and can be practiced from October to February; in addition, hunters may ask for and obtain an exceptional authorisation to trap and kill Egyptian mongooses between March and May. In Spain, this mongoose is considered a pest by hunters, because of its presumed impact on small game species. Despite these threats, the Egyptian mongoose has a wide distribution, so it is unlikely to become seriously endangered in the foreseeable future.

GENUS *ICHNEUMIA*

There is only species in this genus, the white-tailed mongoose, which lives in Africa and is the largest mongoose in the world.

WHITE-TAILED MONGOOSE

Ichneumia albicauda

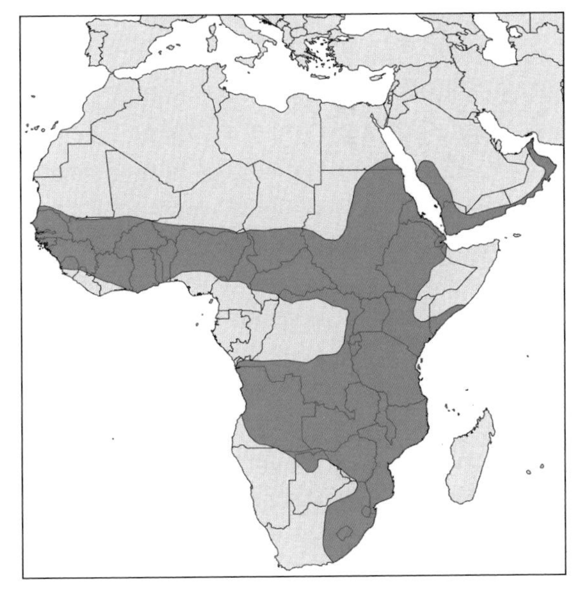

Total length: 91–151 cm
Weight: 3.5–5 kg
Red List status: Least Concern

White-tailed mongooses are a grizzled grey, with long blackish legs, and as the common name implies, most of the bushy tail is white; however, some individuals are very dark grey to black, and have a black tail. The tip of the muzzle can be dusky brown and the cheeks whitish, and the ears are fairly large and round. The feet have five digits, with strong, curved claws.

White-tailed mongooses are found across Africa, south of the Sahara, and also along the southern edge of the Arabian Peninsula. They live in woodlands, savannah, scrub, grasslands, and also in farmland, plantations, and around towns and villages; they do not occur in swamps, dense rainforests, and deserts.

White-tailed mongooses are mostly solitary. The size of their home ranges is dependent on the availability of food and dens and can be up to around 4 km^2, with males having fairly exclusive ranges that overlap those of two or three females, and females have ranges that often overlap. In areas where there is a high density of females, several related females (usually mothers and daughters) may share a home range.

White-tailed mongoose (South Africa © Lourens Swanepoel)

White-tailed mongooses are mainly nocturnal, and are particularly active during the first third of the night and on darker nights. During the day, they rest in termite mounds, amongst rocks, in disused buildings, at the base of trees, and in burrows dug by other animals, changing den sites from day to day. Females sometimes share their daytime rest sites with juveniles or other females – there is little social interaction, and individuals generally just tolerate each other.

The diet of white-tailed mongooses is mainly insects, especially termites, beetles, grasshoppers, and crickets; however, they are opportunistic and will eat other invertebrates (including earthworms, millipedes, scorpions, and spiders) and small vertebrates (small mammals, reptiles, amphibians, and birds), as well as some fruit and carrion. They feed predominantly on termites and ants in the dry season, and dung beetles in the rainy season. When foraging, white-tailed mongooses use a rapid zig-zag trot, with occasional brief stops. They can catch several insects a minute and may forage for six to eight hours a night, travelling 4–5 kilometres. Occasionally, white-tailed mongooses dig for insects, and will pounce on small vertebrate prey. They can kill poisonous snakes, and eat eggs by hurling them between their back legs to smash open on rocks. Several individuals may forage together when large swarms of insects are emerging, but they do not socialise in any way. In towns and villages, white-tailed mongooses are attracted to the cloud of insects that form below street lights, as well as the crawling insects on dung piles in cattle enclosures, and may also raid garbage cans and other human refuse sites. They will also hunt insects on roads, and are sometimes hit by vehicles.

Although white-tailed mongooses are largely silent, they will mutter while digging for insects, and if threatened, may growl, grunt, or bark. Individuals will scent mark their home range with anal gland secretions, urine, and dung. They deposit scats at their den sites and at central middens that may be used by several neighbouring adults, and will urinate by characteristically arching their tail. The tail may also be used in threat displays – a threatened white-tailed mongoose will erect its fur and tail, as well as vocalise.

White-tailed mongooses are seasonal breeders and the birth of pups appears to coincide with wet periods in several regions: March to April and October to December in

East Africa, and October to February in southern Africa; however, the dry season may be the favoured breeding period in West Africa, from January to February. During the mating period, a male and female may copulate for several hours. A female will give usually give birth to one or two pups, but sometimes three or four, which she rears alone. Young males are more likely to disperse when they are independent, whereas females may remain on their mother's home range, which can lead to the formation of female clusters or clans.

Some large carnivores, such as leopards and African wild dogs, may kill white-tailed mongooses, but their obnoxious anal gland secretions, coupled with an impressive threat display, may deter many predators. In captivity, white-tailed mongooses can live up to 14 years.

White-tailed mongooses have a wide geographical distribution and can live in a variety of habitats, including human inhabited areas, which suggest that this is a robust species that is not particularly threatened, although individuals are killed on roads, during local predator control operations, and by dogs in rural areas.

GENUS *LIBERIICTIS*

There is only one species in this genus, the Liberian mongoose, which lives in West Africa. Remarkably, this mongoose species was only discovered by Western scientists in 1958, when a German anthropologist obtained eight skulls from villagers in northeastern Liberia, and it was only in 1974 that two complete specimens were obtained by American zoologists. The first live animal, a single male, was captured in 1989 in Liberia, and was transported to Toronto Zoo, in Canada. In the Ivory Coast, Liberian mongooses were observed for the first time in Taï National Park in 1990, and one male was trapped and radio-collared in 1999, which helped facilitate regular observations of three groups over a three-year period. Recent camera-trapping surveys have confirmed that Liberian mongooses are present in Sapo National Park, in southeastern Liberia.

LIBERIAN MONGOOSE

Liberiictis kuhni

Total length: 63–67 cm
Weight: 2–2.5 kg
Red List status: Vulnerable

Liberian mongooses have a dark brown coat, a pale throat, and a dark stripe, bordered by faint white ones, on each side of the neck. The head is elongated, with a long snout, and the ears are small and round. The legs are darker toward the extremities, with feet that

Liberian mongoose (© Kevin Schafer)

have five digits (although the first one is small) and long, curved claws. The tail is bushy, gradually tapering towards the tip.

Liberian mongooses are found in a small region of West Africa (Liberia and Ivory Coast), and live in primary and secondary evergreen forests, particularly in swamp forest, along sandy streambeds, and in forested areas that have moist soils. This is a social species, with typically groups of four to six animals living together, although larger groups have been observed. Adult males are often observed alone, and one radio-collared male frequently travelled between three groups, joining them for one to three days at a time.

Liberian mongooses are diurnal, and a group sleeps together at night in hollow logs, under fallen trees, or occasionally in termite mounds, and usually use a different den on consecutive nights. They eat mainly earthworms, but may also feed on insect larvae, small vertebrates, and fruit. Liberian mongooses forage in riverine wetland areas and near streambeds with deep sandy soils, where they dig for earthworms and other soil invertebrates by first using the front feet to excavate the earth, before sticking their muzzle into the soil. Liberian mongooses can also climb palm trees to forage for worms or beetle larvae. Each group returns to the same area to forage approximately once every three to four weeks, and their digging activities may result in considerable disturbance of the soil.

Liberian mongooses communicate with each other using soft grunting sounds. They are often found in close association with sooty mangabeys. Both species are preyed upon by crowned hawk-eagles, and the mongooses respond to the monkeys' anti-predator warning calls by quickly dispersing and running into thick vegetation, or under fallen trees. An aggressive encounter was once observed with a group of cusimanses: four

Liberian mongooses initiated the encounter and fought off a larger group of ten common cusimanses by advancing towards them and giving threatening growls.

Virtually nothing is known about their breeding in the wild, except that the birth of pups may occur in the middle of the rainy season, which lasts from May to September, and when invertebrates are probably most available.

As this is a forest species, it is likely to be severely threatened by deforestation across West Africa, resulting from agricultural, logging, and mining activities. Even though Liberian mongooses are found in secondary forests (which are forests that have regrown after timber harvesting), we do not know how much they are impacted by forest disturbances, or whether they can tolerate any harmful effects resulting from tree extractions over the long term. The Liberian mongoose is heavily hunted for bushmeat throughout its range, using dogs, shotguns and snares, and this is undoubtedly reducing local numbers, but since this is not closely monitored, it is not clear how severe this threat may be to the survival of this species in some areas. Liberian mongooses may also be vulnerable to the heavy use of pesticides in forest plantations, as worms are known to accumulate toxins at levels dangerous to mammalian predators.

GENUS *MUNGOS*

There are two species in this genus, the Gambian mongoose and banded mongoose, which both live in Africa. Whereas the banded mongoose is one of the better known and more intensively studied mongoose species, and is a popular animal in zoos and television programmes, the Gambian mongoose is a very poorly known, unstudied species, with very little information about its natural history and conservation status in the wild.

GAMBIAN MONGOOSE

Mungos gambianus

Total length: 53–74 cm
Weight: 1–2.2 kg
Red List status: Least Concern

Gambian mongooses have coarse, grizzled, brownish-grey fur, with a yellowish throat and chest, and a distinctive brownish-black streak on both sides of the neck that runs from the ear to the foreleg. They have a bushy tail that tapers to a black tip, and five digits on each foot, with very long claws, particularly on the forefeet.

Gambian mongoose (© Ondřej Machač)

Gambian mongooses are found in West Africa, and live in dry to semi-moist woodlands, open savannah, grassland, and coastal scrub. This is a social species, with family groups of up to 40 individuals living together, although 5–15 is more typical. Gambian mongooses are active during the day, and sleep together at night in termite mounds, and perhaps in abandoned burrows. They feed mainly on invertebrates, and some small vertebrates, including rodents, snakes, and lizards. Snails and eggs are thrown against something hard, either horizontally backwards through the hind legs or vertically down to the ground. Each member of the group twitters continuously while foraging. Virtually nothing is known about their breeding behaviour, except that young animals were seen during January, February and September in Ghana, and at the end of June in Sierra Leone.

Although we know very little about this mongoose species, it is listed by the IUCN as Least Concern, as it is believed to be common throughout its range and not thought to face any major threats. However, Gambian mongooses are hunted for bushmeat, and are sometimes considered a pest by farmers, who may possibly persecute them in some areas. Like so many other mongoose species, we need to learn more about its natural history, and the precise impacts of any human-induced threats, before we can more reliably assess its conservation status and extinction risk.

BANDED MONGOOSE

Mungos mungo

Total length: 50–67 cm
Weight: 0.9–1.9 kg
Red List status: Least Concern

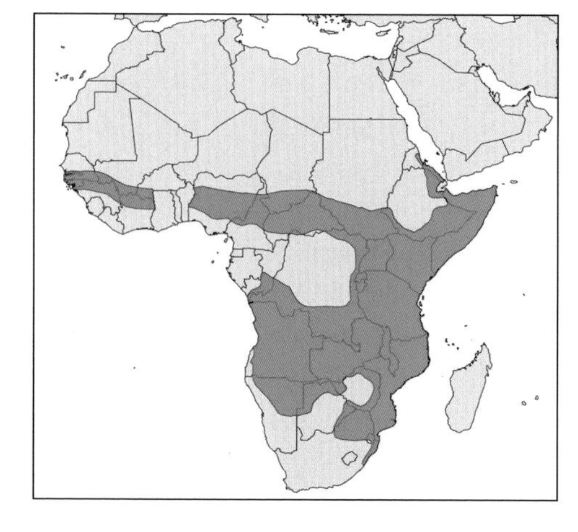

Also known as striped mongoose

Banded mongoose (Botswana © Laurent Vallotton)

Banded mongooses are grizzled pale grey-brown to dark brown, with 10–15 dark, narrow, transverse bands across the back, from the shoulders to the base of the tail. They have a long head, with a pointed muzzle, and small, rounded ears. They have five digits on their feet, although the first one is smaller than the rest, and each has very long, curved claws.

Banded mongooses are found across most of Africa, south of the Sahara, and mainly live in savannah, woodland, and grassland. They can also be seen around towns and villages, but are absent from deserts, semi-deserts, and montane regions. Banded mongooses live in social groups of up to as many as 75 individuals, although around 15–20 is more normal, with a group containing adult males and females, and their young. Each group shares a territory that can be up to a couple of square kilometres in size, and may move several kilometres in a day when foraging for food. The size of a group and its territory depends on the type of habitat and the availability of food, and this varies geographically. All the members of a group cooperate in rearing young and repelling predators, and will even care for sick and injured individuals. There is no conspicuous dominance hierarchy, except during the breeding season, when males will compete for females. For most of the time, group members are not aggressive towards each other, although fights over food may break out, with the finder or owner usually being the winner.

Banded mongooses are active during the day, emerging in the early morning and returning before sunset, and will sleep together at night in termite mounds, in the hollow of tree roots, underneath rock falls, and in burrows that are often located in gullies or thickets, and which have multiple entrances and chambers. A group can have up to 40 dens within its territory that are occupied for just a few days at a time, or are used for longer periods when there are young pups. During the day, a group forages together, usually resting in a shady area around midday. Banded mongooses also spend time scent marking their territory and watching out for predators, as well as engaging in social play, and grooming both themselves and other group members – several individuals may also groom a warthog to remove ticks and lice that they will eat.

Banded mongooses are largely terrestrial, but they can climb to the tops of termite mounds, which they use as vantage points, and can also climb trees if chased by a terrestrial predator, such as an African wild dog.

The diet of banded mongooses is mainly insects, such as beetles, termites and crickets, as well as other invertebrates, including snails, millipedes, and spiders. Occasionally they catch small vertebrates, such as mice, rats, frogs, lizards and small snakes, and will also eat bird eggs and fruits, as well as feeding in garbage dumps in villages. When foraging, banded mongooses move together as a group, but each individual will fan out to seek food on its own, sniffing and scraping at the ground, and often stopping and digging intensively for live prey. Adult mongooses will aggressively defend both their foraging area and any food items that they find, unless they are providing food for young pups. The dung piles of large herbivores, such as elephants, are popular sites for foraging, as they are usually covered by crawling beetles and other insects. Banded mongooses can crack open hard-shelled food, such as a large insect or egg, by throwing it backwards between their hind legs onto a rock or other hard surface. Millipedes, frogs and other prey that produce noxious secretions are rolled around in the earth before being eaten.

Since banded mongooses are social, group-living animals, it is not too surprising to learn that they communicate with others using a wide variety of sounds and calls. While busy looking for food, each mongoose almost continuously emits a low, grunting 'contact' call that helps to keep the group together, and if an individual finds itself separated from the rest of the group, it will utter a distress 'lost' call. To defend a foraging site, a banded mongoose will growl and spit at an approaching pack member. 'Lead' calls are used by individuals to entice the whole group to follow, and specific 'pup-follow' calls encourage pups to follow an adult. Shrill, chirruping 'war' cries alert group members to a rival group, and also encourages them to charge at the invaders. 'Worry' calls warn group members of a low risk danger, whereas short, sharp alarm calls elicit rapid evasive behaviour – banded mongooses will even respond to the alarm calls of some other bird species.

Banded mongooses have excellent vision and can distinguish bird predators from other non-predatory birds at a great distance. When a distant predator is sighted, the group will immediately vocalise and stand upright to gain a better view of the danger. If a raptor or large terrestrial predator, such as a leopard, then approaches too closely, pack members will give a high-pitched alarm call that sends the entire group running for cover. However, banded mongooses can drive off smaller terrestrial predators, such as jackals or servals, by bunching tightly together and advancing slowly, whilst vocalising aggressively.

Scents and smells are also used to communicate between group members and rival groups. Pack members will scent mark each other, and also objects within their territory, such as rocks, by wiping them with the sticky, pungent secretions from their anal gland – this is often a communal event, with all the group members involved in an orgy of scent marking. Individuals that become lost during foraging trips can relocate their group by scent, and upon reintroduction to the pack, the returning individuals are vigorously marked by the other group members. Males can determine the reproductive status of females in neighbouring groups by monitoring the scent marks at shared marking posts, and if oestrous females are detected, the males from one group will make deep forays into the neighbouring group's territory in search of extra-group copulations. Depositing urine and faeces in conspicuous places is another way in which banded mongooses leave scent and visual signals for rival groups.

When two banded mongoose groups come across each other, the smaller one will usually retreat, without any physical aggression, but skirmishes can occur, and some individuals have been killed during such encounters. Upon sighting each other, group members will stand erect and give a distinctive, screeching call that alerts the rest of their pack to the presence of a rival group. A large group will usually cause a smaller one to flee, with members of the larger group chasing behind. Groups that are more closely matched in size will bunch together and approach each other with caution, stopping frequently to stand upright and stare at their opponents. Individuals of both groups will then approach to within twenty or thirty metres of each other before rushing forward and fanning out to engage in one-to-one fights, or chases, with their opponents. Although face-to-face confrontations between individuals may only last a few seconds, they are often violent, involving repeated bites and scratches with the front paws, until one of the combatants bolts. Groups that have become scattered will sometimes retreat, then bunch together and advance again – in this way, fights between evenly-matched groups can sometimes last for over an hour.

Breeding occurs during the wetter periods in regions that have a marked seasonality, otherwise banded mongooses may breed at any time of the year. Females in dry regions have one or two litters a year, whereas in wetter equatorial regions, they may have up to five. All the females in a group enter oestrus at around the same time, and during this period, males exhibit an obvious dominance hierarchy when competing for access to females, with larger males generally becoming the most dominant. In contrast, females do not compete or interfere with each other for matings, although older females may enter oestrus a few days earlier, and mate sooner, than younger females, which results in older females having larger litter sizes. The most dominant males will guard oestrous females for two to three days, and aggressively repel subordinate males who attempt to sneak copulations. However, females go to considerable lengths to escape their guarding male in order to mate with other males within the group, and both males and females will also seek copulations with individuals in neighbouring groups – males and females will even lead their group deep into a neighbouring group's territory in pursuit of additional copulations. The consequence of all this mating activity is that although the dominant males sire most pups, almost all the males and females in a group manage to mate with numerous partners, and about three quarters of the females will eventually give birth.

The gestation period lasts approximately nine weeks, and females may give birth to up to six pups in an underground den, although three is most common. These births are synchronised, with many females giving birth on exactly the same day – those born on different days are less likely to survive. Pups are born blind and with short fur, and weigh only 20–50g. Their eyes open at around ten days, and the pups stay in an underground den until they are weaned, at around three to four weeks of age. During this period, pups may be transferred between dens, two or three times. Whilst the pups are in the den, one or more adults will babysit them while the rest of the group goes off to forage. The babysitters guard the communal litter against predators, such as snakes and monitor lizards, and even against rival groups of banded mongooses, which may sometimes attack a natal den and kill the pups inside. The babysitters usually change every day, with dominant males babysitting more than subordinates, and adult females and sub adults the least. At around four weeks old, pups will start to accompany adults on short afternoon foraging trips, and a week or so later, they will leave the den in the morning with the rest of the group. For the next couple of months, each pup will have a personal escort while outside the den, a particular adult that remains in charge of the pup until it is independent. Each pup closely follows its escort around, fending off other pups that approach too closely, and will beg for food, which the escort provides by dropping whole or partially-eaten prey items. Adults in a group will defend pups against predators, and pups will often shelter under the belly of their escort when frightened or threatened. Adults will also carry pups to the next den, and groom and play with them.

Groups of young males and females, between one and three years old, will sometimes voluntarily leave their natal group, or be aggressively driven out by the other pack members, especially during the oestrus period. These dispersing groups may either invade and take over another established group, driving out the same-sex residents, or will attempt to establish their own territory, which may lead to aggressive fights with resident groups.

Banded mongooses are vulnerable to a wide array of predators, including marabou storks, martial eagles, monitor lizards, and leopards. Pup mortality is particularly high, and at least half the pups may die before reaching independence. Banded mongooses can live to over ten years in the wild, and 17 years in captivity.

Banded mongooses are hunted for bushmeat, which may reduce their numbers in some areas, but since they have a wide distribution across Africa, are able to live close to humans, and occur in numerous protected areas, it seems unlikely that banded mongooses will be threatened in the foreseeable future.

GENUS *PARACYNICTIS*

There is only species in this genus, Selous' mongoose, which lives in Africa.

Selous' mongoose

Paracynictis selousi

Total length: 67–91 cm
Weight: 1.4–2.2 kg
Red List status: Least Concern

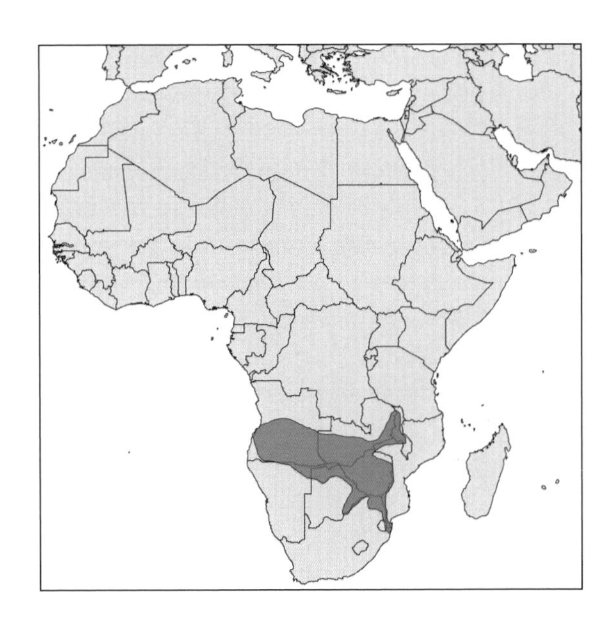

Selous' mongooses are grizzled pale to tawny-grey, with dark brown or black lower legs and feet, and a darkish tail that has a white tip. The head has a pointed muzzle and rounded ears that are partially covered in the front by long hairs. There are four digits on each foot, with long, slightly curved claws.

Selous' mongooses are found in central-southern Africa and live in savannah grassland and woodland, particularly in areas that are dominated by sandy substrates, and are absent from dense forests, deserts, and semi-deserts.

Selous' mongooses are mainly solitary, but are sometimes seen in pairs, which are likely a mating male and female, or a mother with her young. They are mainly nocturnal, and will rest during the day in burrows that they can excavate themselves if the ground is sandy, or they will use burrows that have been dug by other species. These burrows can have several passages and chambers, and may have one or two entrances that emerge under the shelter of a low bush, or directly out into the open. When threatened, Selous' mongooses may bolt into any available hole, including those of springhares and aardvarks. Like meerkats, they can stand up on their hind legs and hold their head high to look for danger.

Their diet is mainly insects, especially termites, beetles, locusts and grasshoppers, but also includes other invertebrates, such as spiders and scorpions, as well as small rodents, lizards, snakes, amphibians, and birds. They will readily dig for insects in the ground

Selous' mongoose (Phinda Nature Reserve, South Africa © Gonçalo Curveira Santos)

and scratch for prey among leaf litter and at the bases of grassy tufts. When searching for food, they move with their head low and nostrils close to the ground, which suggests that mongooses can locate their prey by smell. It is also believed that they have good hearing, which may help them find subterranean food.

Very little is known about the breeding behaviour of Selous' mongooses, but it seems that births mainly occur during the warm, wet summer months, from August to March, and the litter size can be up to four pups. The only known predator is the martial eagle.

The IUCN considers Selous' mongoose to be uncommon across its range, but believes that its habitat is not at high risk and there are no major threats, although this species might be hunted for bushmeat, and dogs and cats (both domestic and feral) may out-compete or even prey on Selous' mongooses in some areas. Clearly, field studies are urgently needed to provide more ecological information for better conservation assessments.

GENUS *RHYNCHOGALE*

There is only one species in this genus, Meller's mongoose, which lives in Africa.

MELLER'S MONGOOSE

Rhynchogale melleri

Total length: 72–91cm
Weight: 1.8–2.8 kg
Red List status: Least Concern

Meller's mongooses are grizzled pale to dark brownish-grey, with a paler head and darker lower limbs, and a bushy tail that can be predominately black, brown, or greyish-white. The muzzle is quite blunt and appears slightly swollen. There are five digits on each foot, with the first digit extremely reduced (particularly on the hind foot), and each has short, curved, and sharp claws.

Meller's mongooses are found in eastern Africa, and live in savannah, open woodland and grassland, especially in areas with termite mounds, and also occur in montane bamboo forest, up to at least 1,850 metres. They are nocturnal, solitary and mainly eat termites, but occasionally they will feed on grasshoppers, beetles and centipedes, and also small vertebrates, including small snakes and frogs, as well as some fruits. Litters of up to three pups are born in a burrow, most likely during the wet summer season, between November and January.

Meller's mongooses appear to be uncommon, and may have a patchy distribution. The IUCN considers that this species is unthreatened, as it is present in several protected

Meller's mongoose (South Africa © Lourens Swanepoel)

areas, its preferred habitat is extensive across eastern Africa, and there seem to be no obvious major threats. However, miombo woodlands across south-central Africa are being cleared for agriculture, or are heavily disturbed through timber extraction, and Meller's mongooses might be hunted for bushmeat. There has also been considerable expansion of the human population in Tanzania and Zambia, and their domestic dogs may kill Meller's mongooses.

GENUS *SURICATA*

The meerkat is the only species in this genus, and it lives in Africa. This is undoubtedly the best known and most intensely studied of the mongoose species, appearing in many documentaries, films and TV shows, and exhibited in numerous zoos around the world.

MEERKAT

Suricata suricatta

Also known as suricate, and slender-tailed meerkat

Total length: 45–53 cm
Weight: 0.6–1 kg
Red List status: Least Concern

Meerkats have pale-greyish tan to yellow-brown fur, with short dark brown streaks across the back, black eye patches on the face, and a slender tail that has a black tip. The small,

Meerkat (South Africa © Emmanuel Do Linh San)

rounded ears can be closed to keep dust out while digging, and each foot has four digits, with very long claws.

Meerkats are found in southern Africa, and live in open semi-arid deserts, dry open savannah, dry scrubland and grassland, and are absent from true deserts, dense forests, and mountainous terrain. Meerkats live in family groups that comprise a dominant breeding pair, related and unrelated adults of both sexes, and offspring from several litters. As many as 30–50 meerkats can live in a group, but troops of around twelve individuals are much more typical. Meerkats are highly social, with each mongoose helping to guard and protect its group from predators, and to raise young, as well as grooming and scent marking each other to maintain a group bond – meerkats will even care for injured members. Each group shares a territory of up to 10 km^2, which is defended by the whole pack. When two meerkat groups encounter each other, they each perform a 'war dance' – adopting a tiptoe body stance, and erecting their body hairs and tails. Chases and fights may then break out, which can sometimes end with some individuals being severely wounded, or even killed.

Meerkats are active during the day, and a group sleeps together overnight in an underground den. Meerkats are good diggers and can excavate burrows themselves, but they often occupy dens that have been dug by other small mammals, and will sometimes share a burrow with other species, including springhares, ground squirrels, yellow mongooses, Cape grey mongooses and common slender mongooses, apparently without any hostilities. Each burrow can have multiple entrances, extensive passages and several chambers, and temperatures inside are more stable and comfortable than the hot and cold extremes on the surface. A meerkat group may have around five dens within their territory that are each used for a few days at a time, although particular dens may be used for longer periods when pups are around. Groups also have many boltholes on their territory that they run to when danger threatens, and meerkats are seldom more than 50 metres from the nearest safe haven.

Meerkats emerge from their underground burrow in the early morning. They usually start the day by sunbathing, and then head out to forage for food. They move along as a tight group, with individuals rarely more than five metres apart, and will forage for five to eight hours, often resting in a shady area around midday. A group may cover a distance of up to 6 kilometres each day, before returning to the den around sunset.

Meerkats mainly eat invertebrates, especially beetles, termites, scorpions, spiders, millipedes, centipedes, and various insect eggs, pupae and larvae, but occasionally they will feed on small vertebrates, such as rodents, lizards, geckos, frogs, snakes, birds, and eggs. Their diet varies depending on prey availability, which is affected by the type of habitat in which they forage, as well as the weather. Each meerkat searches for food on its own, by sniffing and scraping at the ground surface, and digging intensively for prey – sometimes even disappearing from view in the deep hole that they have dug. A meerkat will defend its own foraging patch from other group members, except for young individuals that are still dependent on adults for providing food. Fights over captive prey are common, but these are usually settled in favour of the owner.

Meerkats have several physiological and behavioural adaptations that enable them to survive in a hot, semi-arid environment. Their metabolic rate is only 58% of that found in other mammals of a similar size, which reduces the amount of internal heat produced. Meerkats can also lose heat through evaporation without losing too much water, and at air temperatures as high as 40°C, they can resist overheating for up to five hours by panting like a dog. Meerkats also avoid extreme high midday temperatures by taking siestas, and they often sprawl on damp ground to cool themselves down. As daily temperatures decrease and the day length shortens, meerkats start foraging later and finish their activities before sunset, and will spread themselves over warm rocks. Meerkats avoid the problem of rapid heat loss at night by sheltering in burrows and huddling in groups for extra warmth; huddling is particularly important for pups, which are vulnerable to the cold.

Meerkats are small animals that live in open habitats, which makes them very vulnerable to predators, especially when they have their noses to the ground, busily looking for food. So while a group forages, some members of the pack will act as sentinels, using prominent features, such as the tops of large termite mounds, as look-out posts for scanning both the sky and horizon. To get a better view of their surroundings, meerkats will often stand on their hind legs using the tail as a prop or fifth limb. Before going 'on guard', a meerkat will give a 'watchman's song' to inform the group that it is on sentry duty. Although these guard duties are shared by all the group members, the amount of time an individual will devote to this task depends on its status within the pack; for instance, older subordinate males spend more time guarding than other pack members.

Meerkats communicate with each other, and rival groups, using a wide variety of calls and sounds, as well as scents and smells. While searching for food, meerkats keep in constant touch with each other by using a 'contact' call. An individual will defend its food patch by growling and spitting at any approaching group member, and pups will give 'begging' and 'give me food' calls, when an adult finds a food item. 'Worry' calls warn group members of a low-risk danger, while a short, sharp bark from a sentinel will send the colony scattering to their burrow. Pack members will wipe the pungent secretions from their anal glands both on themselves and on raised objects, such as rocks, as a means of stamping the group's identity on each other and their territory. This activity is often done communally, with all group members involved in synchronous bouts of scent marking. Territory borders and burrow entrances are frequently scent marked, and this is mainly done by the dominant pair in the group. Urine and faeces are also used for scent marking, with many individuals in a group using the same latrine sites.

Female meerkats are at least two years old before they can conceive. The birth of pups is often timed for the wet summer months, when invertebrate food is more likely to be abundant, but it can occur at any time of the year. Since mating is rarely observed above ground, it probably occurs more frequently in the underground den. There is a dominance hierarchy amongst both sexes, and usually only the dominant pair breeds. An alpha male might aggressively prohibit subordinate males from mating, and subordinate females may be prevented from conceiving through physiological changes in hormone levels, as well as being evicted by the alpha female during the late stages of her pregnancy (although subordinate females are allowed to return to the group after the birth of her pups). However, these reproductive suppression mechanisms are not 100% infallible, and some subordinate males and females do manage to have successful matings, so that overall, an alpha female may produce 75% of the litters within a group, and the alpha male fathers 80% of the pups.

The gestation period is around 60 to 70 days, and up to eight pups are born within a litter. The number of litters per year can be up to four, with an alpha female producing more litters than subordinates. If females other than the alpha female give birth within a group, all the births usually occur within one week of each other. The pups are born with short hair and their ears closed, and weigh only 25–36g at birth; their eyes are also closed at birth, and will open at around 10–14 days. The pups of subordinate females are often killed by the alpha female, and pregnant subordinates may also kill pups, including those of other subordinates and the dominant female's. Pups will suckle from numerous females, including those that have not given birth, and will be weaned at around two months of age.

The pups stay in the underground den until they are three to four weeks of age. During this time, one or more adults will babysit the pups, while the rest of the group goes out and forages. These babysitters usually stay at the burrow for the whole day, feeding little or not at all, and their role is to keep the pups warm and safe from predators. Breeding adults do relatively little babysitting, while non-breeding subordinates, or 'helpers', do most of the work. Pups begin accompanying the foraging group when they are around four weeks old. Outside the den, helpers will provide pups with food and defend them against predators.

Unlike the banded mongoose, there is no personal pup escorting system, and meerkat pups will beg for food from any adult, by calling continuously and intensely. Helpers vary widely in how much of the food they will share with pups, and the amount given is gradually reduced as pups approach independence, at around three months of age. Young meerkats reach adult weight when they are around one year old, and may live up to 15 years in the wild and around 20 years in captivity.

Some individuals may remain in their natal group for all, or much, of their lives, whereas some yearlings and adults may disperse to join other groups, with males more likely to leave than females. Dispersal of adults often happens early in the breeding season, when hostility within a group is at its greatest. Males usually leave a group voluntarily, whereas subordinate females rarely do, and instead, may be forcibly expelled from the group by dominant females during the latter weeks of their pregnancy, although many return soon after the birth of the pups. Few emigrants successfully establish new groups, and although old males may be ousted by a gang of males taking over the group, male immigrants may sometimes have to wait for days to gain acceptance to a new group.

Meerkats are killed by snakes and raptors, including Cape cobras, martial eagles, tawny eagles, lanner falcons, and pale chanting goshawks. Other larger carnivores, such as lions, spotted hyaenas and black-backed jackals, are also a potential threat. However, meerkats will bunch together to mob many predators, such as cobras, depending on the size of the group; more than five meerkats will mob a jackal, whereas a smaller group will more likely flee to a safe haven.

The meerkat is not considered threatened by the IUCN, as it is believed to be a common species that is relatively widespread across southern Africa, and is present in several protected areas. There is currently no evidence that infection with tuberculosis, which may be common in this species, has led to any direct persecution in farming areas, or that the current small trade in meerkats as pets has significantly affected wild populations. Meerkats have been killed in some areas during rabies control efforts, even though these were mainly targeted at the yellow mongoose.

Genus *Urva*

There are nine Asian species in this genus: small Indian mongoose, short-tailed mongoose, Indian grey mongoose, Indian brown mongoose, Javan mongoose, collared mongoose, ruddy mongoose, crab-eating mongoose, and stripe-necked mongoose. They were all originally included in the genus *Herpestes*, but recent genetic studies have shown that they should now be placed in the genus *Urva*. However, they may still be listed under *Herpestes* in other publications.

The small Indian mongoose and the Javan mongoose were previously considered the same species by some authors, and separate ones by others, but recent genetic studies have shown that they should be treated as two separate species. The past confusion over the taxonomic status of these two mongooses has meant that most of the published ecological information that was supposedly about the Javan mongoose, which lives in Southeast

Asia, actually relates to the small Indian mongoose, which is found in the Indian region. So in reality, we actually know very little about the Javan mongoose, and most field studies on the small Indian mongoose have been done on islands where it was introduced.

A subspecies of the small Indian mongoose (*Urva auropunctata palustris*), which occurs in Bengal, is advocated by some taxonomists to be another species, called the Bengal mongoose or marsh mongoose, but there is currently insufficient evidence to warrant this.

A species called Hose's mongoose was described by a naturalist in the 1800s, using just one specimen that was supposedly collected from the island of Borneo. This museum specimen is very similar to the short-tailed mongoose, but has a slightly differently-shaped jawbone and a more reddish coat. A recent genetic study showed that this mongoose is not a valid species, and that the original specimen is just a slightly odd short-tailed mongoose.

The mongoose species that lives on Palawan, in the Philippines, was generally considered to be the short-tailed mongoose, which also occurs in other places in Southeast Asia. A recent genetic study showed that the Palawan mongoose is, in fact, the collared mongoose.

SMALL INDIAN MONGOOSE

Urva auropunctata

Total length: 38–74 cm
Weight: 0.5–1 kg
Red List status: Least Concern

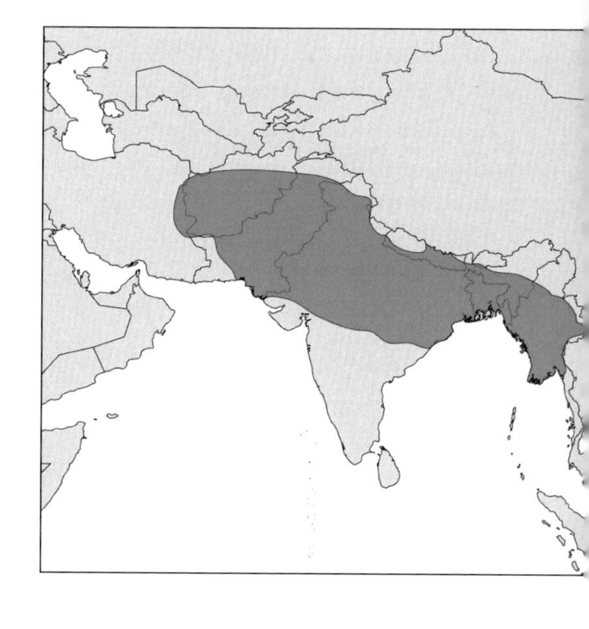

Small Indian mongooses are buff to rufous or dark yellowish-grey, and the hairs have white and dark bands that give an overall grizzled appearance. The muzzle is pointed, and the ears are short. The tail is muscular at the base and tapers throughout its length. There are five digits on each foot, with sharp non-retractile claws.

The native range of the small Indian mongoose extends from the Middle East, across northern India, to southern China, where it lives in open forests and scrub, and sometimes close to human habitations. Small Indian mongooses were introduced (mainly for rodent and snake control in agricultural areas) to over 60 islands in the Pacific and Indian oceans, and the Caribbean and Adriatic seas, as well as mainland Europe (Croatia, Bosnia and Herzegovina, Montenegro) and South America (Guyana and Suriname); a failed attempt was made in Australia, and a few small Indian mongooses sighted on mainland Florida, in North America, were captured through intensive trapping. A couple of individuals were also trapped on the Pacific island of New Caledonia, but no others have been seen since.

Small Indian mongoose (India © Manuel Ruedi)

Most of the information we know about this species comes from field studies in places where it has been introduced. Small Indian mongooses are active during the day and they will rest and sleep at night amongst fallen trees, in holes within tree root systems, and in burrows. They are usually solitary and occupy home ranges of up to around one square kilometre in size, with a variable amount of overlap of ranges, depending on the location. In some places where there are abundant food resources, a few males mongooses may form small social coalitions.

Small Indian mongooses have a wide diet that varies according to the locality and season. They eat invertebrates, including insects, centipedes and scorpions, as well as small vertebrates, such as mice, rats, lizards, snakes, toads, and birds, along with some eggs and fruit. To kill centipedes and scorpions, small Indian mongooses bite them repeatedly and toss them aside, whereas to kill rodents, birds and snakes, they drive their canines into the brain or vertebral column.

Small Indian mongooses scent mark objects within their home range by wiping them with the secretions from their anal glands, and they are able to distinguish the scent marks of other individuals. For a solitary species, small Indian mongooses have a large vocal repertoire of at least 12 distinct calls, including a squawk, spit, bark, and growl. They can dig vigorously and climb trees, but are rarely seen far above the ground.

The timing of reproduction may be related to day length, as most pregnancies seem to occur prior to the summer solstice. About every three weeks, a female is receptive, or in oestrus, for around three to four days. A female may have two to three litters a year. At the beginning of oestrus, a female is restlessness and scent marks more often. Several males may gather around a female, and they commonly scream, bark, and chase each other. A mating pair may copulate several times a day. The pregnancy period, or gestation, is around 49 days, and up to five pups may be born in a litter. Newborns weight about 21g, and are only covered with light grey hairs. The incisors and the eruptive cones of the canines are visible, and the claws are well developed. The eyes are closed, and will open around 17 days later. Young pups will mew if they are disturbed. At two weeks, the incisors are fully in place and the canines have erupted, and at 22 weeks, all the permanent teeth are in place. Pups take their first excursion out of the den at about four weeks, and will begin following their mother on hunting trips at six weeks. At around four months, the young are two-thirds of the adult body mass, and will reach sexual maturity at one year.

The small Indian mongoose is not considered by the IUCN to be endangered in its native range, as it can live in many different types of habitat, even close to humans, so habitat modifications may not significantly affect this species. However, large numbers of mongooses are hunted in India for their hairs, which are used for making paintbrushes, and small Indian mongooses are often captured and sold as pets in India, Nepal, and China.

Unfortunately, introduced populations of small Indian mongooses on oceanic islands have helped cause the extinction of several species of rare and endemic birds, mammals and reptiles, and they eat the eggs and young of some endangered sea turtles. There have been some attempts to eliminate small Indian mongooses from where they are not native, but only on a few small islands have they been eradicated.

SHORT-TAILED MONGOOSE

Urva brachyura

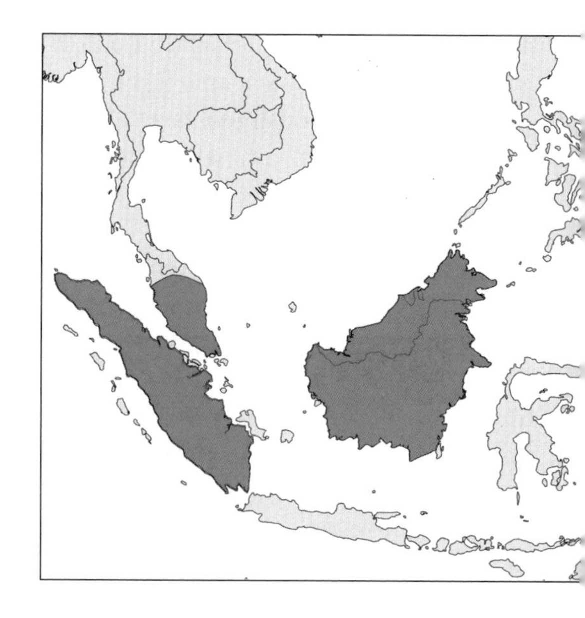

Total length: 54–74 cm
Weight: 2–3 kg
Red List status: Near Threatened

Short-tailed mongooses are a grizzled dark brown. The eyes are reddish-brown, the ears are small, and the nose is reddish-orange. The tail is quite short, usually around half the head and body length, and tapers from the base to the tip, appearing conical in shape. There are five digits on each foot (although the first one is smaller), with quite long claws.

Short-tailed mongoose (Malaysia © Jayasilan Mohd-Azlan)

Short-tailed mongooses are found across Southeast Asia, and mainly live in lowland evergreen rainforest, often close to rivers and small streams, but may sometimes occur in farmland adjacent to forests if there is sufficient ground cover. Short-tailed mongooses are mainly solitary, although two individuals – possibly a mating pair or a mother and her young – are sometimes seen together. Short-tailed mongooses are active during the day, although they often rest for short periods, especially around midday. At night they may sleep in a hollow log. Males have home ranges that can be up to 2.5 km^2 in size, whereas those of females are smaller, up to 1.5 km^2. Females have exclusive ranges, and a male's home range encompasses at least one or two females. Not much is known about the diet of short-tailed mongooses, except that they feed on invertebrates, including insects, spiders and crabs, and are reported to eat small vertebrates, eggs, and fruits. Short-tailed mongooses are mostly terrestrial, but there is some evidence that they can climb trees, at least up to one metre off the ground.

Short-tailed mongooses are primarily a lowland forest species, and rainforests across its range are being cleared at an alarming rate, which is undoubtedly having severe impacts on this species. Even though short-tailed mongooses have sometimes been recorded in some degraded habitats, such as logged forests and farmland, it is not known how well this species can tolerate habitat disturbances, and there is no evidence that it can survive in non-forested habitat that is remote from native pristine forest. Short-tailed mongooses are often found near watercourses, so water pollution could pose a serious threat, and even though there is no evidence that hunters specifically target short-tailed mongooses, they are captured as a by-catch in snares that are set for other animal species.

Indian grey mongoose

Urva edwardsii

Also known as common grey mongoose

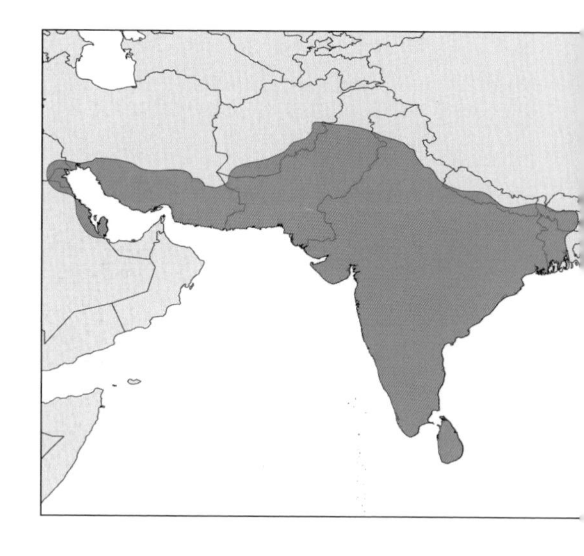

Total length: 68–90 cm
Weight: 1–2 kg
Red List status: Least Concern

Indian grey mongooses are a grizzled tawny-grey or pale grey; some individuals, particularly from northern areas, have a reddish colouration, especially on the muzzle, legs, and tail. The lower legs are usually darker than the body, ranging from rufous-brown to black, and the tail is long, with a whitish or yellowish tip. There are five digits on each foot.

Indian grey mongooses are found in India and neighbouring countries, on Sri Lanka, and in small isolated pockets across the Middle East. This mongoose was introduced to Italy and Peninsular Malaysia (and possibly Sumatra), but these attempts failed, and this species no longer exists in these countries.

Indian grey mongooses live in dry forests, particularly in open shrubby areas, and also occur in plantations and near human settlements. They are mainly solitary, but mating

Indian grey mongoose (India © Manuel Ruedi)

pairs and females with young have often been seen. An adult male radio-tracked for three months in India had an overall home range of 0.2 km^2. Indian grey mongooses are active during the day, and rest and sleep under rocks, bushes, and in holes at the base of trees or in the ground. Their diet include small vertebrates, including rats, mice, birds, snakes, lizards, and frogs, and a wide variety of invertebrates, such as insects, scorpions, and centipedes, as well as eggs, fruits, human refuse, and carrion.

Indian grey mongooses are an aggressive predator of snakes, and are commonly kept in India and Pakistan for staged fights with cobras. They will adopt a defensive posture when attacked by rolling into a ball, placing their head between their legs and under the bushy tail, and erecting the long hairs on the back and tail to shield the body.

Indian grey mongooses may have two to three litters in a year, with births mainly occurring during May to June and October to December. The gestation period is 56–68 days, and two to four pups are born in a litter. The pups are helpless and blind at birth, but develop rapidly, and will remain with their mother for up to six months if she does not mate again.

The Indian grey mongoose is considered by the IUCN to be a common species throughout its range, and able to adapt to human-dominated landscapes. However, large numbers of mongooses are hunted in India for their hairs, which are used for making paintbrushes, shaving brushes and good luck charms, as well as to supply meat for local people. Indian grey mongooses are often captured and sold as pets, and some tribes from northern India hunt them for their skins, which they sell in local markets in Nepal. Indian grey mongooses are also captured by local tribes in Pakistan for staged fights with cobras.

INDIAN BROWN MONGOOSE

Urva fusca

Also known as brown mongoose

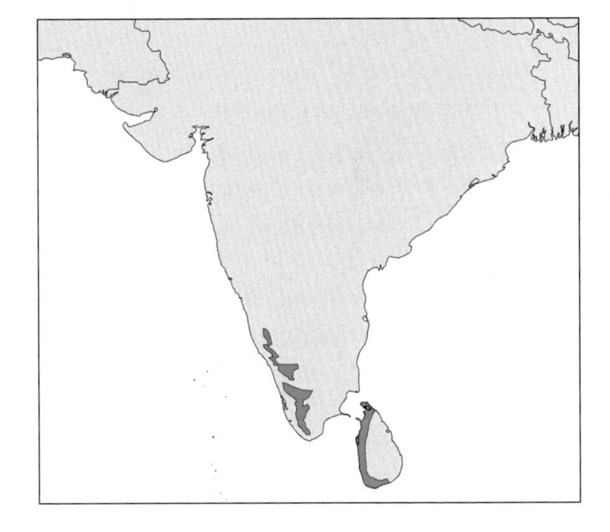

Total length: 53–82 cm
Weight: 1.9–2.7 kg
Red List status: Least Concern

Indian brown mongooses are a grizzled dark brown, with a slightly paler head, and tawny chin and throat. The feet are almost black, and the tail is bushy and conical. They are found in southwest India and western Sri Lanka, and live in dense rainforests, but can also occur in adjacent areas, including tea and coffee plantations. Indian brown mongooses are mainly nocturnal and solitary, but occasionally pairs or small family groups are seen. Nothing is known about their diet,

Indian brown mongoose (India © Divya Mudappa)

but it is assumed that they feed on a variety of invertebrates and small vertebrates, and they are known to scavenge on the carcasses of large mammals, and to feed in rubbish dumps. Indian brown mongooses may breed in burrows beneath rocks and tree roots, and to have three to four young.

The Indian brown mongoose has a restricted range, and was previously thought to be rare, but because there has been a recent increase in sightings and camera trap records, it is now considered by the IUCN to be locally common. Preliminary radio-tracking of one individual showed surprisingly large movements for an animal of its body size, which may suggest that population densities are low (or that home ranges are much overlapping). Although specific threats to this species are not well known, the destruction of forests is very likely having severe impacts on populations. And even though Indian brown mongooses have been recorded in plantations, there is no current evidence that they can persist within a modified landscape, far from native forest. Large numbers of mongooses are hunted in India for their hairs, which are used for making paintbrushes, shaving brushes and good luck charms.

A population of Indian brown mongooses has been discovered on the island of Viti Levu in Fiji (on which there is also the introduced small Indian mongoose), which may have derived from a pair that escaped from a local private zoo in the late 1970s.

JAVAN MONGOOSE

Urva javanica

Also known as small Asian mongoose

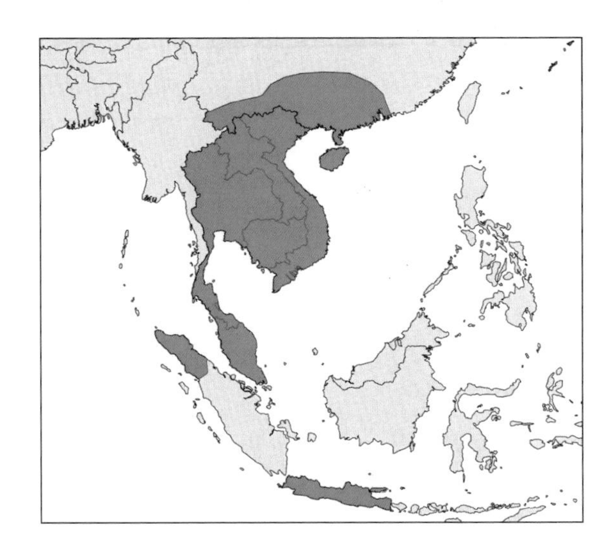

Total length: 51–73 cm
Weight: 0.5–1 kg
Red List status: Least Concern

Javan mongooses are grizzled buff to dark brown, often with a reddish-brown head, and the legs are the same colour as the body or slightly darker. The muzzle is pointed, the nose is blackish and the eyes and ears are small. The tail tapers throughout its length. There are five digits on each foot, with quite long, sharp, non-retractile claws.

Javan mongooses are found across Southeast Asia. Their presence on the island of Sumatra is unclear, in that they have only been recorded in the north, and it is currently not known if Javan mongooses occur across the whole of Sumatra (and have simply been overlooked by field scientists), or if they were introduced to a few localities on the island.

Javan mongoose (Java, Indonesia © Manuel Ruedi)

Javan mongooses live in rainforests, scrubland, grasslands, and in open areas, including rice fields and farmland. They are solitary and active during the day, and may sleep in holes in the ground or at the base of a tree. Their diet is reported to include rats, birds, reptiles, frogs, crabs, and insects. Javan mongooses may breed throughout the year, and produce litters of two to four pups, after a gestation period of about six weeks.

The Javan mongoose is considered to be a widespread and common species that apparently can survive in human-modified landscapes, although the status of populations in the wild is unknown. Javan mongooses are hunted and sold for meat in local markets in

China, Laos, Thailand and Vietnam, and they are also captured and traded for food, or as a novelty pet, in the wildlife markets of Medan, in North Sumatra, and in Jakarta, on Java.

Collared mongoose

Urva semitorquata

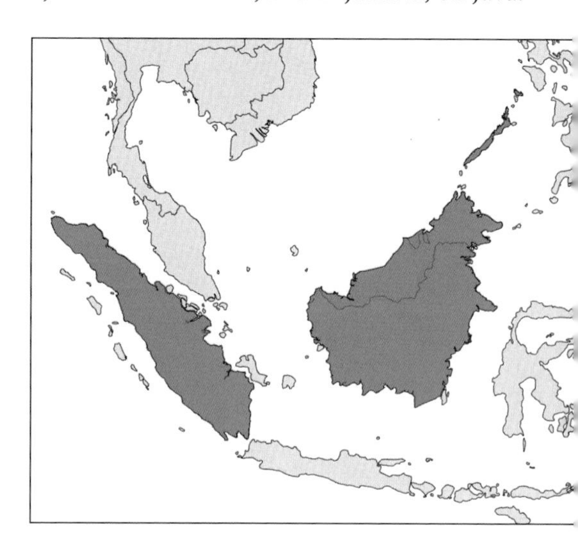

Total length: 66–76 cm
Weight: 2–4 kg
Red List status: Near Threatened

Collared mongooses are dark reddish-brown, or sometimes orange-brown. The underside of the head, throat and chest are yellowish, and a whitish stripe runs along the neck. The lower parts of the legs are blackish-brown. The tail is yellowish, and is relatively longer than in the short-tailed mongoose.

Collared mongooses are found on Borneo and Sumatra; however, this species has only been recorded in a few locations on Sumatra, and further information is needed to understand its status there. A recent genetics study discovered that the mongooses on Palawan Island, in the Philippines, are in fact collared mongooses, and not short-tailed mongooses, as was previously thought.

Collared mongooses live in primary and logged lowland rainforest, peat swamp forest, mixed secondary-acacia forest mosaics and, occasionally, in plantations. They are solitary

Collared mongoose (Borneo, Malaysia © Andrew James Hearn and Joanna Ross)

and mainly diurnal. Virtually nothing is known about their diet, except that it includes small animals and ants.

The fast disappearance of rainforests throughout Southeast Asia is undoubtedly having severe impacts on this species. Even though collared mongooses have been recorded in somewhat degraded habitats, such as logged forests, it is not known how well this species can tolerate forest disturbances, and whether it can persist in heavily-modified landscapes that include substantial areas of oil palm plantations and farmland. Although there is no indication that collared mongooses are specifically hunted or targeted for the wildlife trade, incidental captures, using unselective trapping methods such as snares, are reducing numbers in some areas.

RUDDY MONGOOSE

Urva smithii

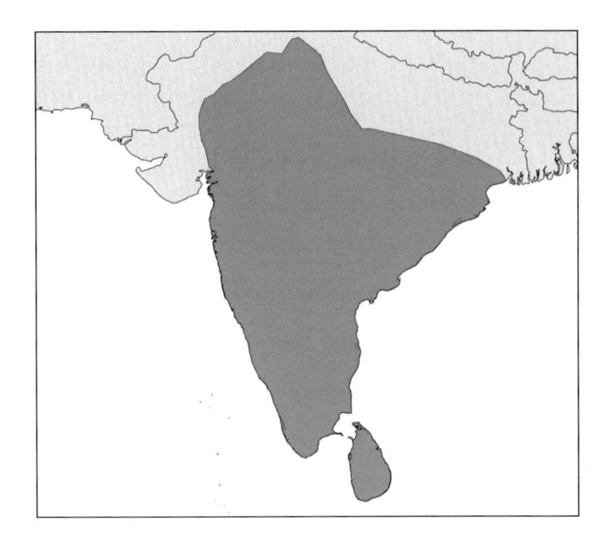

Total length: 74–94 cm
Weight: 1.8–2.7 kg
Red List status: Least Concern

Ruddy mongooses are a grizzled greyish-brown, with a tinge of rufous on the underparts, and the long tail has a black-tasselled tip. The lower limbs and feet are darker than the rest of the body, with partially webbed digits.

Ruddy mongoose (India © Kalyan Varma)

Ruddy mongooses are found across India and Sri Lanka, and live in dry forests, thorn scrub, and dry grassland-forest mosaics. They also occur in disturbed forest, but avoid heavily modified habitats near humans. Very little is known about their diet, but they have been seen feeding on mice, snakes and birds, including doves, partridges and quails, and may also scavenge, especially on road-killed carcasses. Ruddy mongooses are mainly active during the day and are usually solitary, although mating pairs and family groups of up to five animals have been seen. They are mainly terrestrial, but may occasionally climb trees.

The IUCN does not consider the ruddy mongoose to be an endangered species since it has a wide distribution, can occur in some human-modified habitats and is believed to be common in some parts of India. However, the destruction of forests may be having an impact on populations, and large numbers of mongooses are hunted in India for their hairs (which are used for making paint brushes), and for supplying meat to local people.

CRAB-EATING MONGOOSE

Urva urva

Total length: 71–90 cm
Weight: 3–4 kg
Red List status: Least Concern

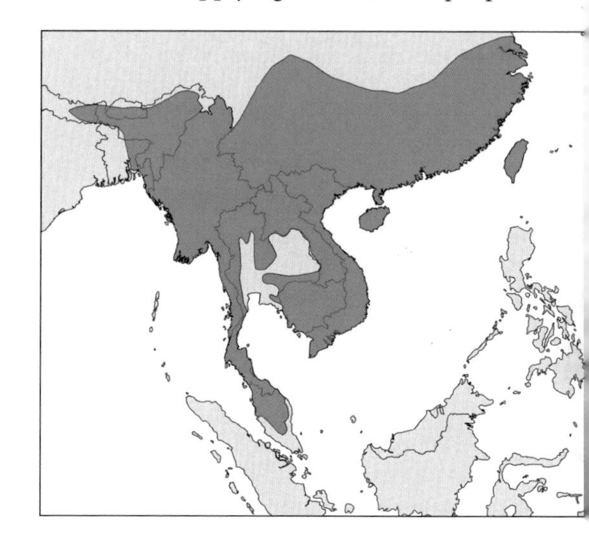

Crab-eating mongooses have a grizzled grey to dark brown shaggy coat, and a distinctive white stripe that runs along the side of the neck from the cheek to the shoulder. The top of the head is pale greyish-brown and the chin and throat are whitish. The muzzle is pale yellowish, with a flesh-coloured nose, and the ears are short and rounded. The limbs are brown to black, with feet that have strong claws, and the bushy tail becomes progressively tawny or whitish towards the tip. There are five digits on each foot.

Crab-eating mongooses are found in parts of South Asia, Southeast Asia and China, and live in evergreen and deciduous forest, especially in wetlands, small marshes, scrubby areas and along streams, up to at least 2,000 metres; they also sometimes occur in rice fields and other agricultural areas. Crab-eating mongooses are active during the day, and will rest and sleep in holes in the ground and amongst rock crevices. They are mainly solitary, although family groups of up to four have been seen, and occupy home ranges that can be at least 1 km^2 in size.

Crab-eating mongooses eat a wide variety of invertebrates, including insects, crabs, earthworms, spiders and centipedes, as well as snakes, frogs, small mammals, birds, fish, and some fruit. Their diet varies with changes in the availability of food in different

Crab-eating mongoose (India © Vijay Anand Ismavel)

habitats. They hunt along the banks of streams, feeling under stones and in rock crevices, and scratch, dig, and sniff at the ground. They are good swimmers and do not hesitate to enter water. Crabs and snails are taken with the forepaws and either lifted and smashed on a rock or thrown between the hind limbs onto a hard surface behind. It is uncertain when crab-eating mongooses may breed, but the gestation period is 50–63 days, and two to four pups are born in a litter.

In the northern part of its range, including Laos, Vietnam, Cambodia and China, the crab-eating mongoose is hunted for its meat and fur, and live animals are sold as pets in Cambodia. Despite these hunting pressures, the crab-eating mongoose is not considered threatened by the IUCN, as it is found in a wide variety of habitats, including degraded and fragmented areas, and is believed to be resilient to the heavy hunting that is occurring in large parts of its range. However, the destruction of forests is very likely causing declines in populations.

STRIPE-NECKED MONGOOSE

Urva vitticollis

Total length: 65–87 cm
Weight: 2.2–3.4 kg
Red List status: Least Concern

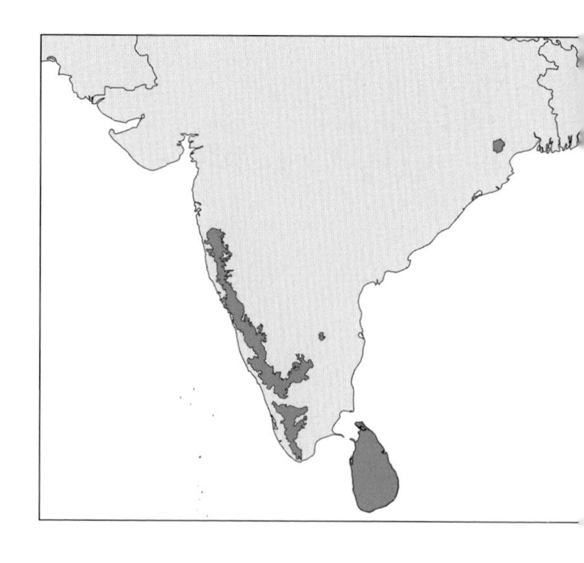

Stripe-necked mongooses are a grizzled grey-brown on the forequarters, and a grizzled rufous-brown on the hindquarters; some individuals are entirely rufous, except for the head, which is always steely-grey. A distinctive black stripe, with a white border, runs from behind the ear to the shoulder. The lower limbs are dark purplish-brown, with five digits on each foot, and the orange-red tail has a black-tasselled tip.

Stripe-necked mongooses are found in southwest India and on Sri Lanka, and live in evergreen and deciduous forests, especially in swampy clearings and along watercourses, and also occur in tea and teak plantations, rice fields, and open scrubland. Stripe-necked mongooses are diurnal and are usually solitary, except during the mating season or when a female has young. The diet includes small mammals, birds, reptiles, insects and eggs, and it is likely that they eat crabs, frogs and fish, as they often hunt for food along the banks of rivers and in swampy areas; they may also scavenge on carcasses.

Female stripe-necked mongooses may give birth to two or three pups in a litter. On Sri Lanka, a female was observed suckling three young on a dry patch of earth, under an overhanging mass of rocks.

Stripe-necked mongoose (India © Yathinsk, CC BY-SA 3.0, Wikimedia commons)

The stripe-necked mongoose has a restricted range and appears to be uncommon. A chief threat to this species may be habitat loss, particularly the destruction of forests, as well as hunting. The stripe-necked mongoose is also regularly killed by dogs, and sometimes in retaliation for raids on poultry farms.

Genus *Xenogale*

There is only one species in this genus, the long-nosed mongoose, which lives in Africa. This mongoose has sometimes been included in the *Herpestes* genus, but recent genetic studies have supported the placement of this species in *Xenogale*.

Long-nosed mongoose

Xenogale naso
Also known as long-snouted mongoose

Total length: 72–103 cm
Weight: 2–4.5 kg
Red List status: Least Concern

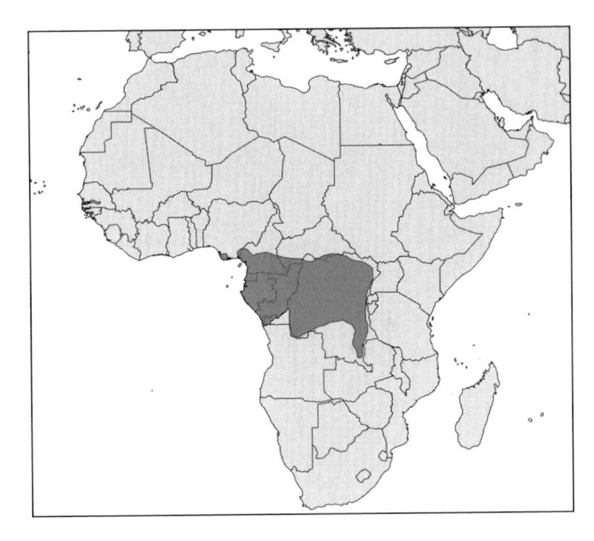

Long-nosed mongooses are a grizzled brownish-black, with a long muzzle and prominent black nose. The head is greyish, with ears that are round, broad and low set, and the tapered tail has long black hairs. The limbs are deep brown to black, and the feet have five digits, quite long claws, and are partly webbed.

Long-nosed mongooses are found in central Africa and live in rainforests, near streams, on streambeds and in swampy areas with dense and tangled understorey. They avoid open forest, but sometime forage on burnt grassland close to forest edges. Long-nosed mongooses are mostly solitary, and their home range can be up to 0.5 km^2 in size. They are mainly active during the day, and may sleep at night in a hollow log. They mainly eat insects, including beetles, termites, locusts and ants, and other invertebrates, such as snails, millipedes, spiders and scorpions, but also feed on small mammals, frogs, toads, snakes, lizards, skinks and birds, as well as crustaceans, fish, fruits, and berries. Long-nosed mongooses break open snails by throwing them backwards through the hind legs against a tree or a rock.

Very little is known about the breeding behaviour of long-nosed mongooses, but young animals have been seen in March in West Africa, with a litter size of at least three pups. The young reach adult size at around seven months of age and permanent dentition is attained at one year. Leopards are known to kill long-nosed mongooses, and the remains of a long-nosed mongoose was found in the scat of the black-legged mongoose, but this

Long-nosed mongoose (Gabon © Laila Bahaa-el-din)

may have resulted from scavenging. One long-nosed mongoose lived up to 11 years in captivity.

The long-nosed mongoose is not considered by the IUCN to be endangered, as it is relatively widespread and believed to be abundant in some areas. However, long-nosed mongooses may be threatened by the loss and fragmentation of forests due to logging, mining and slash-and-burn farming, and they are also hunted for their meat.

APPENDICES

WHERE TO SEE MONGOOSES

Most mongoose species are solitary and shy of humans, which makes it very difficult to see them in the wild. This is particularly true of those species which are active only at night or which live in densely vegetated habitats. If you are able to visit an African or Asian rainforest, though, you might be lucky enough to see a mongoose moving around during the day, or if you slowly walk along a trail or logging road at night and scan the undergrowth with a powerful torch or spotlight, you might detect the bright, reflective eyeshine of a nocturnal mongoose.

The best chance of seeing mongooses in the wild is offered by the group-living, social mongoose species that live in open habitats, and this includes the meerkat, banded mongoose, and common dwarf mongoose. These are all common species and active during the day, foraging in groups in open, dry savannah, which makes them easier to spot. They all live in sub-Saharan Africa, and the national parks and reserves are good places to see these social mongooses. For example, banded and common dwarf mongooses can be seen in Kruger National Park, in northeastern South Africa. You can visit habituated groups of meerkats in various places across southern Africa: Botswana's Makgadikgadi Pans National Park is one of the best places to see them, and Tswalu Kalahari Reserve in South Africa is another good spot. Although difficult to access, one place that has a high diversity of mongooses, up to eight species, is the Chinko Nature Reserve in eastern Central African Republic.

Seeing mongooses in a zoo is perhaps a more realistic opportunity for most people to view these fascinating animals. Many zoos around the world house mongooses, particularly some of the social species, such as the meerkat, banded mongoose, and common dwarf mongoose. A search on the internet can identify which zoos have certain species. We would encourage you to visit those zoos that house animals in good conditions, have good education programmes and are actively involved in conservation, with captive breeding programmes and supporting field research.

CLASSIFICATION OF MONGOOSES

The following list is the classification of the mongooses presented in this book. There is still some debate about the status of some species, particularly in the genera *Bdeogale* and *Galerella*. Species can be divided into subspecies, and numerous subspecies have been proposed for most of the mongooses, but these require much more investigation and confirmation, particularly using modern molecular techniques, and so we have not included them here.

Family: Herpestidae
Subfamily: Mungotinae

Genus *Crossarchus*

- Alexander's cusimanse (*Crossarchus alexandri*)
- Ansorge's cusimanse (*Crossarchus ansorgei*)
- Common cusimanse (*Crossarchus obscurus*)
- Flat-headed cusimanse (*Crossarchus platycephalus*)

Genus *Dologale*

- Pousargues' mongoose (*Dologale dybowskii*)

Genus *Helogale*

- Somali dwarf mongoose (*Helogale hirtula*)
- Common dwarf mongoose (*Helogale parvula*)

Genus *Liberiictis*

- Liberian mongoose (*Liberiictis kuhni*)

Genus *Mungos*

- Gambian mongoose (*Mungos gambianus*)
- Banded mongoose (*Mungos mungo*)

Genus *Suricata*

- Meerkat (*Suricata suricatta*)

Subfamily: Herpestinae

Genus *Atilax*

- Marsh mongoose (*Atilax paludinosus*)

Genus *Bdeogale*

- Bushy-tailed mongoose (*Bdeogale crassicauda*)
- Jackson's mongoose (*Bdeogale jacksoni*)
- Black-legged mongoose (*Bdeogale nigripes*)

Genus *Cynictis*

- Yellow mongoose (*Cynictis penicillata*)

Genus *Galerella*

- Kaokoveld slender mongoose (*Galerella flavescens*)
- Somali slender mongoose (*Galerella ochracea*)
- Cape grey mongoose (*Galerella pulverulenta*)
- Common slender mongoose (*Galerella sanguinea*)

Genus *Herpestes*

- Egyptian mongoose (*Herpestes ichneumon*)

Genus *Ichneumia*

- White-tailed mongoose (*Ichneumia albicauda*)

Genus *Paracynictis*

- Selous' mongoose (*Paracynictis selousi*)

Genus *Rhynchogale*

- Meller's mongoose (*Rhynchogale melleri*)

Genus *Urva*

- Small Indian mongoose (*Urva auropunctata*)
- Short-tailed mongoose (*Urva brachyura*)
- Indian grey mongoose (*Urva edwardsii*)
- Indian brown mongoose (*Urva fusca*)
- Javan mongoose (*Urva javanica*)
- Collared mongoose (*Urva semitorquata*)
- Ruddy mongoose (*Urva smithii*)
- Crab-eating mongoose (*Urva urva*)
- Stripe-necked mongoose (*Urva vitticollis*)

Genus *Xenogale*

- Long-nosed mongoose (*Xenogale naso*)

Web links for the sources of pictures

Page 3: Picture sources: https://en.wikipedia.org/wiki/Carnivora
Page 8 Left: Licence information: https://creativecommons.org/licenses/by-sa/3.0
Picture source: https://commons.wikimedia.org/wiki/File:Black_mongoose_waterberg.jpg
Page 18 Top Right: Licence information: https://creativecommons.org/licenses/by-sa/2.0
Picture source: https://commons.wikimedia.org/wiki/File:Skeleton_of_a_Meerkat.jpg
Page 18 Bottom: Licence information: https://creativecommons.org/licenses/by-sa/3.0
Picture source: https://commons.wikimedia.org/wiki/File:White-tailed_mongoose_(Ichneumia_albicauda).JPG
Page 35 Right: Licence information: https://creativecommons.org/licenses/by-sa/3.0
Picture source: https://commons.wikimedia.org/wiki/File:Meerkat_in_Namibia.jpg
Page 45: Picture source: https://commons.wikimedia.org/wiki/File:Cynictis_penicillata_mating2.jpg
Page 46: Licence information: https://creativecommons.org/licenses/by-sa/3.0
Picture source: https://en.wikipedia.org/wiki/Indian_grey_mongoose#/media/File:Baby_Mongooses.jpg
Page 58 Left: Licence information: https://creativecommons.org/licenses/by-sa/2.0/fr/deed.en
Picture source: https://commons.wikimedia.org/w/index.php?curid=19499288
Page 58 Center: Picture source: https://commons.wikimedia.org/w/index.php?curid=18787378.
Page 58 Right: Licence information: https://creativecommons.org/licenses/by-sa/2.0/fr/deed.en
Picture source: https://commons.wikimedia.org/w/index.php?curid=19499288
Page 59 Left: Licence information: https://creativecommons.org/licenses/by-sa/3.0
Picture source: https://commons.wikimedia.org/wiki/File:SAMA_Kubera_1.jpg
Page 59 Right: Picture source: https://commons.wikimedia.org/wiki/File:2007_0811collection-Bertsch0123.JPG
Page 60: Picture source: http://www.philamuseum.org/collections/634-549.html#object/88340
Page 61 Left: Picture source: https://commons.wikimedia.org/wiki/File:JunglebookCover.jpg
Page 61 Right: Picture source: https://commons.wikimedia.org/wiki/File:Jungle_book_p206r.png
Page 73 Left: Licence information: https://creativecommons.org/licenses/by-sa/2.0
Picture source: https://upload.wikimedia.org/wikipedia/commons/f/f4/Myanmar_Illicit_Endangered_Wildlife_Market_04.jpg
Page 74: Licence information: https://creativecommons.org/licenses/by-sa/3.0
Picture source: https://upload.wikimedia.org/wikipedia/commons/e/e4/Bushmeat_-_Buschfleisch_Ghana.JPG
Page 99: Licence information: https://creativecommons.org/licenses/by-sa/3.0
Picture source: https://commons.wikimedia.org/w/index.php?curid=10723312
Page 142: Licence information: https://creativecommons.org/licenses/by-sa/3.0
Picture source: https://commons.wikimedia.org/w/index.php?curid=21281822

FURTHER READING

Estes, R.D. (2012) *The Behavior Guide to African Mammals: Including Hoofed Mammals, Carnivores, Primates*. University of California Press.

Francis, C.M. (2008) *A Field Guide to the Mammals of South-East Asia*. New Holland Publishers.

Gilchrist, J.S., Jennings, A.P., Veron, G. & Cavallini, P. (2009) Family *Herpestidae*. In: *Handbook of the Mammals of the World, Volume 1, Carnivores*, D. Wilson & R. A. Mittermeier (Eds). Lynx Edicions, pp. 262–328.

Hunter, L. & Barrett, P. (2018) *A Field Guide to Carnivores of the World (Second Edition)*. Bloomsbury Wildlife.

Kingdon, J. & Hoffmann, M. (Eds). (2013) *Mammals of Africa. Volume V: Carnivores, Pangolins, Equids and Rhinoceroses*. Bloomsbury Publishing.

Lewis, S. & Llewellyn-Jones, L. (2018) *Animals in Antiquity: A Sourcebook with Commentaries*. Routledge Publishers.

Phillipps, Q. & Phillipps, K. (2016) *Phillipps' Field Guide to the Mammals of Borneo*. Princeton University Press.

Prater, S. H. (1990) *The Book of Indian Animals (Third Edition Reprint)*. Oxford University Press.

Shepherd, C.R. & Shepherd, L.A. (2012) *A Naturalist's Guide to the Mammals of South-East Asia*. John Beaufoy Publishing.

Smith, A.T. & Xie, Y. (2008) *A Guide to the Mammals of China*. Princeton University Press.

Stuart, C. & Stuart, M. (2017) *Stuarts' Field Guide to the Larger Mammals of Africa (Fourth Edition)*. Struik Nature Publishers.

The IUCN Small Carnivore Specialist Group produces a journal, *Small Carnivore Conservation*, which publishes articles on small carnivores, including mongooses, and can be accessed online: <www. smallcarnivoreconservation.org>

Taxonomic index